For leigh
Four cats
Forever

Memory Practices in the Sciences

Inside Technology
edited by Wiebe E. Bijker, W. Bernard Carlson, and Trevor Pinch

A list of books in the series appears at the back of the book.

Memory Practices in the Sciences

Geoffrey C. Bowker

The MIT Press
Cambridge, Massachusetts
London, England

First MIT Press paperback edition, 2008

This book was set in Baskerville by SNP Best-set Typesetter Ltd., Hong Kong.

Library of Congress Cataloging-in-Publication Data

Bowker, Geoffrey C.
Memory practices in the sciences / Geoffrey C. Bowker.
p. cm.—(Inside technology)
Includes bibliographical references and index.
ISBN 978-0-262-02589-8 (hc. : alk. paper)—978-0-262-52489-6 (pb. : alk. paper)
1. Knowledge representation (Information theory) 2. Knowledge, Theory of.
3. Science—Philosophy. 4. Science—Information technology. I. Title. II. Series.
Q387.B69 2005 501—dc22 2005050492

Contents

Acknowledgments

Diogenes, les voyant en telle ferveur mesnaige remuer et n'estant par les magistratz employé à chose aulcune faire, contempla par queluqes jours leur contenence sans mot dire. Puys . . . feit hors la ville tirant vers la Cranie . . . une belle esplanade, y roulla le tonneau fictil, qui pour maison luy estoit contre les injures du ciel, et, en grande vehemence d'esprit desployant ses braz, le tournoit, viroit, brouilloit, barbouilloit, hersoit, versoit, renversoit, nattoit, grattoit, flattoit, barattoit, bastoit, boutoit, butoit, tabustoit, cullebutoit, trepoit, trempoit, tapoit, timpoit, estouppoit, destouppoit, detraquoit, triquotoit, tripotoit, chapotoit, croulloit, elançoit, chamailloit, bransloit, esbransloit, levoit, lavoit, clavoit, entravoit, bracquoit, bricquoit, blocquoit, tracassoit, ramassoit, clabossoit, afestoit, affustoit, baffouoit, enclouoit, amadouoit, goildronnoit, mittonnoit, tastonnoit, bimbelotoit, clabossoit, terrassoit, bistorioit, vreloppoit, chaluppoit, charmoit, armoit, gizarmoit, enharnachoit, empennachoit, caparassonnoit, le devalloit de mont à val et præcipitoit par le Cranie, puys de val en mont le rapportoit, comme Sisyphus faict sa pierre: tant que peu s'en faillit qu'il ne le defonçast.

Ce voyant, quelqu'un de ses amis luy demanda quelle cause le mouvoit à son corps, son esprit, son tonneau ainsi tormenter. Auquel respondit le philosophe qu'à aultre office n'estant pour la republicque employé, il en ceste façon son tonneau tempestoit pour, entre ce peuple tant fervent et occupé, n'estre veu seul cessateur et ocieux.

—François Rabelais, *Oeuvres Complètes*

This work spans more years than I care to (or can, to be precise) remember. So let me begin with my appreciation of the Forgotten Interlocutor—your work has become so much a part of me that I cannot find either words to express it or a name to hang the lack of words on: thank you.

However, I do remember with gratitude several classes of experience as well as some distinct discussions.

The classes of experience are, notably, working at the Center for the Sociology of Innovation in Paris, the Graduate School of Library and Information Science at Urbana/Champaign, the Department of Communication

Epigraph: François Rabelais, *Oeuvres complètes* (Paris: Seuil, 1995), p. 522.

at the University of California at San Diego, and now at the Center for Science, Technology and Society at Santa Cruz University. Each of these places, in different ways, provided love, support, and intellectual nourishment. Colleagues, students, and staff in each wove a truly delightful web of enquiry: I am honored to have been part of it and grateful for the opportunity to develop my work in these settings. I would especially like to signal out my new home in at Santa Clara University—the deep commitment of all members of the university to braiding good work with good works has been a source of deep inspiration for me.

A few particulars in no particular order. At Santa Clara University, Don Dodson, Paul Locatelli, John Staudenmaier, Cathy Valerga, Sherill Dale, and Karen Bernosky have made my final work on this book a pleasure. Leigh Star, to whom I have dedicated this book, has as ever been a wonderful friend and colleague—I could not imagine having written this outside our life together. In recent years, conversations with Eliza Slavet and Andy Lakoff about memory have been a joy—even if neither figures in the bibliography. Not to mention Stefan Tanaka, who shares my passion for our own brand of nonlinear history. Katie Vann has been a continual source of inquiry always carried out in a spirit of playfulness. Allan Regenstreif introduced me to Lacan and encouraged me to get lost in Proust. Mike Cole's energy and erudition was a great help. I am constantly grateful to Michel Serres for his exquisite prose and intricate ideas. And to Bruno Latour for his continual string of original insights as well as for his personal generosity. It has always been a pleasure to talk cybernetics with Andy Pickering—may the mangle be with him. David Edge was a wonderful support at different stages through the trajectory of writing this book: his example has always been before me. Charlotte Linde convened a few wonderful memory workshops and encouraged me to range far and wide. John Bowker gave me some wonderful feedback—as well as being a source of inspiration in himself.

I am grateful to the National Science Foundation for its support of my work, notably, in this instance, through a Human and Social Dynamics grant Comparative Interoperability (NSF 0433369) and a Biodiversity and Ecosystem Informatics grant Designing an Infrastructure for Heterogeneity of Ecosystem Data, Collaborators and Organizations (NSF0242241), but also for my Values in Design grant (NSF 0454775). Preparation for the latter made me think long and hard (if not deep) about the design implications of my memory work. I have learned a lot from my collaborators on these grants. Karen Baker's spirit of inquiry has been a source of great inspiration. Helena Karasti, David Ribes, and Florence Millerand have each flavored my work with their own rich visions.

Let me part on a note of sorrow. This is far too short. I really do regret not having produced a list above as voluminous as my influences: I hope that if I have not recognized you above, you can recognize yourself in my work.

Introduction

"Mr Swivett, approaching a facial lividity that would alarm a Physician, were one present, now proclaims, 'Not only did they insult the God-given structure of the Year, they also put us on Catholic Time. French Time. We've been fighting France all our Lives, all our Fathers' Lives, France is the Enemy eternal, —why be rul'd by their Calendar?' "

"Because their Philosophers and ours," explains Mr. Hailstone, "are all in League, with those in other States of Europe, and the Jesuits too, among them possessing Machines, Powders, Rays, Elixirs and such, none less than remarkable, —one, now and then, so daunting that even the Agents of Kings must stay their Hands."

"Time, ye see," says the Landlord, "is the money of Science, isn't it. The Philosophers need a Time, common to all, as Traders do a common Coinage."

"Suggesting as well an Interest, in those Events which would occur in several Parts of the Globe at the same Instant."

—Thomas Pynchon, *Mason & Dixon*

Facts are but the Play-things of lawyers—Tops and Hoops, forever a-spin. . . . Alas, the Historian may indulge no such idle Rotating. History is no Chronology, for that is left to lawyers—nor is it Remembrance, for Remembrance belongs to the People. History can as little pretend to the Veracity of the one, as claim the Power of the other, —her Practitioners, to survive, must soon learn the arts of the quidnunc, spy, and Taproom Wit, —that there may ever continue more than one life-line back into a Past we risk, each day, losing our forebears in forever, —not a Chain of single Links, for one broken Link could lose us All, —rather, a great disorderly Tangle of Lines, long and short, weak and strong, vanishing into the Mnemonick Deep, with only their Destination in common.

The Revd Wicks Cherrycoke, Christ and History.

—Thomas Pynchon, *Mason & Dixon*

Epigraphs: Thomas Pynchon, *Mason & Dixon* (New York: Henry Holt, 1997), p. 192; Thomas Pynchon, *Mason & Dixon* (New York: Henry Holt, 1997), p. 349; Paolo Levi, "The Ravine," in *The Oxford Book of Detective Stories,* ed. P. Craig (Oxford: Oxford University Press), p. 316.

You will find that for every kind of occurrence there are at least three explanations. The most likely, the absolutely certain one . . . and the true one.

—Paolo Levi, "The Ravine"

In the course of human (and nonhuman) history, it is rare enough for a significant new regime of memory practices to develop. M. T. Clanchy (1993) explores one such in England a millennium or so ago, arguing that "the shift from habitually memorizing things to writing them down and keeping records was necessarily prior to the shift from script to print, and was as profound a change in its effects on the individual intellect and on society" (3).

Looking out from the year 1000, then, one can go back to the invention of writing and a subsequent uneven shift to organizational reliance on written records over several thousand years up to the turn of the first millennium after the Christian era. One can also look forward to the propagation of print culture some few hundred years afterward (Eisenstein 1979) and then several centuries after that to the development of the Internet. This book offers a reading of the ways in which information technology in all its forms has become imbricated in the nature and production of knowledge over the past two hundred years.

The starting point will be the Industrial Revolution in England, with the development of new archival forms consequent on the expanded scope of the British state and accompanying new scientific memory practices—for example, in the then central science of geology. The culmination will be a new form of scientific product, the digital database, within a current central scientific arena: biodiversity science.

The story I tell is not a linear, chronological narrative—that artifact of a previous memory regime. My story weaves a path between the Landlord's time and the Reverend Wicks Cherrycoke's "Mnemonick Deep"; between the social and political work of creating an explicit, indexical memory for science and knowledge and the variety of ways in which we continually reconfigure, lose, and regain the past. The interest in the Landlord's expostulation goes beyond its brute equation of time and money. The Landlord is talking about how infrastructures form.

The mnemonic deep. At the extremes sit dance and play, two ways of reading it, and on the plateau wander an infinite number of ways of writing it. One way of reading it is to see ourselves as at any one moment completely able to escape our history, thanks to that little piece of time which is the present, together with motive force, emergence. Hope, desire, creativity, will are projected onto this little piece of time stuff, the present (ever-present, never in reach). This little object, the numinous present, holds our dreams. The past is a thing that you escape at all costs. It has a heavy hand. A hand with a long reach, as many a politician has found—the politician being the concentrated symbol of the person whose past is completely knowable: prurience combined with moral fervor set up a powerful continuing (and ever incomplete) inquisition into the politician's past, the same inquisition that we carry out daily on confessional shows on daytime television with its smorgasbord of choices for redemption. Our past explains who we are; we stand here publicly before you to receive absolution for that past and to follow emergence into a spangly future. . . . The other way of reading it is as a palimpsest, as in Proust's description of Albertine's face as a palimpsest. The infinite faces of the past can be read off the present face. The mnemonic deep lies deep in the eyes of the beloved. We should remember the past and celebrate it. For how else will we savor, texture, explore, adore the real? Evoking the past is a joy and a solace in the present; through it we constitute a narrative ideal present. The timeless present—ever felt more richly, ever receding into the past. For how can we know the past without taking time for it, mortgaging the present to savor the past? In this reading, we learn from the past, seeing the multiple ways it can lead, and we observe ourselves choosing some of those ways (never a single decision; rarely consciously a decision). This numinous present will lead us to the question of money. As Michel Serres (1982) has noted, money is the degree zero of information. It circulates in an ever more ideal space and time (we have gone from gold to silver to base metals to paper to digits) and is exchanged, duly laundered, as something that is without history. Money is the ultimate token of emergence.

Within this metaphorical economy, time as the money of science is constituted of both the mnemonic deep and the numinous present. The time that scientists create has three main features. First is the time of the experiment/field study itself. Scientists play with much longer time scales than most of us (going back billions of years); with much shorter time scales (down to divisions of the nanosecond, to quantum units of time); and with time series and cycles of great complexity—registering, for example, patterns in time series analyses of proxy measures for past climate (tree rings, peat mosses, fossil seeds, astronomical cycles ranging over tens of thousands of years up to millions of years). Agreements about time and timing are fundamental to all science, so a good time standard operates as a gold standard. Second is the time of the scientific enterprise. Much writing in science is historical—opening a scientific paper with an account of the recent relevant history of one's subdiscipline; continuing with what happened in a particular day at a given laboratory. Particular constellations of historiographical stances that are shared among sets of disciplines, or between practitioners of a given discipline.

Finally, there is the ultimate product—the law of nature. All contingency has been removed from the law of nature—it is true over all time and in any place. In the same way, we will see, our globalizing ethnos is without a past and apparent everywhere. The common time that the Landlord refers to is used to create a universe in which the constructed fact is eternally true.[1] In this sense, the past that scientists create can be read as an eternal present.

The institution of the sciences is one of very few modern institutions that claim a perfect memory of the past (law, through precedents, and theology, through heresies, are others). Even tax records decay over time. This book is about the work that goes into creating this avowedly perfect memory—about its textures and discontinuities; about the technologies and techniques that subtend it, and about ways of thinking about it with a view to designing robust scientific databases that contain traces of the past that are currently cast into oblivion. But to get from here to there—or rather from now to then—we will first look at the array of traces of the past that we leave.

What Traces Do We Leave?

In Which It Is Argued that We Leave a Lot of Traces

I rarely think about the traces that I leave in the world as an ecology. I tend to think of them (when at all) quite concretely. First, my library. It operates as a form of external memory for me (when I, rarely, use it) and as a commemoration of things I have read. Its probable fate after my death is its dispersal into a hundred homes. Marginal notes that I have written will lower the selling price rather than attract attention. Second, on the Web. It is interesting to track dead people on the Web. My friends and acquaintances who died before Mosaic are sparsely represented, and when they are it is generally in a classical, canonical academic style (footnote references, bibliographies, etc.). Or in a Mormon database. Those who died more recently carry on a rich afterlife. They often still receive email messages; links to their Web sites rot very slowly; their informal thoughts are often captured on listserv archives, on comments they have left on a Web site (signing the visitors' books). Some people even have "eternal flame" Web sites[2]—where the problem of maintenance is as live as it is for the Olympic torch or the refrigerated truck. Each of these modes of memory was in place before Mosaic, but it is now possible to articulate it

1. B. Latour, B., *Petite réflexion sur le culte moderne des dieux faitiches* (Paris: Les Empecheurs de Penser en Rond, 1996c). This is the double nature that Latour explores.

2. See http://www.venus.co.uk/gordonpask/.

in ways that were previously unworkable. It would take a researcher a lifetime to track down my written traces—where I have signed guest books in weird museums and twee hostels, people with whom I have carried on informal correspondence. Those of us enjoying and being irritated by post-Mosaic syndrome (PMS), leave legible traces across a wide range of our activities in electronic form. Everyone their own Boswell.

When I, rarely, think about the articulation of the set of traces that I am leaving, I have the immediate apprehension that it's not the real me that's out there on the Web. I know the times when I've censored myself (oh problematic concept!) and when I have performed actions to complement—and frequently to confound—a trace. Thus I might write a positive review of a friend's book and then offer close colleagues a different reading.

Taken globally, the set of traces that we leave in the world does without doubt add up to something. It is through operations on sets of traces that I understand an event in which I take part. Tolstoy wrote about the foot soldier in the Napoleonic wars. The soldier he describes cannot have the experience of the war he is waging or the battle he is fighting because the only "global" traces of the war are inscriptions—notably, maps and statistics. There is no scaleable observation that moves from "I was in a copse hiding behind a tree and was terribly confused" to "I took part in Napoleon's bold attack on the left flank." In this case, where is the experience of the war? When we experience a war, we are relying on the aggregations of other experience to ground and shape our experience.

In general, we use scientific representational forms to fashion our experience. Hacking writes about this at the personal level in terms of learning to be a child abuser or a multiple personality by reading the accounts of others (Hacking 1992, 1995). We internalize these accounts and experience them as our own. Žižek claims that there are no pure patients in psychoanalysis now, there are only Jungian patients, Lacanian patients, and so forth: the stuff of our experience (our symptom) is the aggregated experience of others (Žižek 2000). This has always been the case. History was just as multivectorial then as it is now, and our individuality was just as vectored in archives.

With digital archiving in all its forms, a new regime of technologies for holding and shaping experience has emerged. Our past has always been malleable, but now it is malleable with a new viscosity. The new texture of our past is that we can go from the global to the local and back again with great speed. The new analytic objects that emerge are different if we look at the transition of the local through the global (the unit local-global-local, to paraphrase Marx) or the transition of the global through the local (the unit global-local-global). The former will lead into the new, rich interiority that is emerging with faster global exchange of information, people, flora, fauna, and things. We now have so many identities available to us—just geographically I can reasonably lay claim to Celtic, Isle of Man, Australian, French, English, and so forth. The latter leads into a new form of exteriority, in which the map and the statistic are richer than the territory and govern the territory. It is not that we have the ability to aggregate brute numbers—that has been available since the early nineteenth century at least in a number of domains. It is rather that we can aggregate that data along multiple different dimensions and perform complex operations over that set of dimensions. It is the pleats and the folds of our data rather than their number that constitute their texture.

What This Book Is About

The paradox of digits lies in this. The best possible analog representations are produced by digital computers. If you want a flowing sea of lava oozing from a volcano spuming smoke (as doubtless many of us do), then you don't go to the people who produce analog computers. You go binary. The situation is strictly analogous to that of the moving picture we call cinema being constituted of a sequence of still pictures, recreated anew each fraction of a second just like Descartes' discontinuous universe. Does it make any difference that our best apprehensions of data, past viewscapes, and encapsulated memory are brute numbers, binary and static? For they are also folded, fractal, and febrile.

In order to tackle that question, we need to start with the questions of (1) how our personal memories are technically, socially, and formally mediated (local-global-local) as well as in our heads, and (2) how the socionatural world we operate in is produced locally (formally, technically, socially) to be global (global-local-global). Clearly a vast amount has been written about these questions.

In this book, I examine two questions that are fractal subsets of these two:

1. How do scientists figure their own pasts—both as creatures on earth and in terms of disciplinary lineage?
2. How do scientists figure the past of their entities—the earth, the climate, the extinction event?

A central *aporia*[3] that I explore is constituted by the very general condition that what we leave traces of is not the way we were, but a tacit negotiation

3. See the discussion of aporia in G. Agamben Agamben, *Infancy and History: The Destruction of Experience* (London and New York: Verso, 1993). Aporia is a figure whereby the Speaker sheweth that he doubteth, either where to begin for the multitude of matters, or what to do or say in some strange or ambiguous thing (from OED). Etymologically, it derives from a tropos—without path, or without road.

between ourselves and our imagined auditors (whether in the sense of listeners, readers, or moral or economic watchdogs); yet we also need at some level an understanding of what actually happened in order to forge our futures. The aporia takes many forms. When Bill Gates came up against the U.S. government in the antitrust suit against Microsoft, much was made of some internal email correspondence that laid out his company's predatory strategies. This is a standard tale from the early days of email. Now, there are numerous companies that specialize in searching out and destroying all traces of possibly damaging email correspondence; and many organizations have laid down strict rules about what can be said in an electronic conversation. A similar move was made in the 1930s by the Schlumberger company, when it realized that its internal records could be scrutinized by a court—the company shifted very quickly from writing detailed accounts of their practices in French to writing highly sanitized versions in English (Bowker 1994). Similarly, Ed Hutchins (1995, 20) observes that records kept of navigation on navy ships are written with an eye to a future legal enquiry should there be a disaster. Scientific texts are written not to record what actually happened in the laboratory, but to tell the story of an ideal past in which all the protocols were duly followed: the past that is presented should be impregnable—thus perpetual worrying over whether the Millikan oil-drop experiment (he discarded partial charge values for his particles) was fraudulent, whether Pasteur misrepresented his findings or Mendel messed with his peas. It takes a great deal of hard work to erect a past beyond suspicion. When I tell my life story to a boss or a coworker, there are many things that I unmention, discontinuities that I skate over (Linde 1993). It is very rare to commit a story to paper with a view to telling it "wie es eigentlich gewesen ist." Stories are told in a context, under a description (Hacking 1995). The aporia to which we will return is that despite this central fact about record keeping, there is still a need to keep good records. Microsoft Corporation still needs to retain and propagate a memory of how to be a predator; Schlumberger still wants to know how to work around regulations; scientists want to be able to pass on knowledge about how an experiment really works to their students. This brings us centrally to the question of memory practices. Acts of committing to record (such as writing a scientific paper) do not occur in isolation; they are embedded within a range of practices (technical, formal, social) that collectively I define as memory practices. Taken as a loosely articulated whole, these practices allow (to some extent) useful/interesting descriptions of the past to be carried forward into the future.

What Is Memory, that a Person May Practice One?

Memory is often, and wrongly, conceived of as an act of consciousness and associated with what can be called to mind. By this light, it is often seen as the act of deciphering traces from the past. We don't analyze the movements of icebergs by studying the bit that appears above the surface of the sea; nor should we study memory in terms of that which fires a certain set of neurons at a determinate time. We as social and technical creatures engage in a vast span of memory practices, from the entirely non-conscious to the hyperaware.

Consider the total institution. Mary Douglas (1986) argues that "when everything is institutionalized, no history or other storage devices are necessary" (48). If I get processed into a prison, I can survive there as just a number (as the Count of Monte Cristo discovered). There is no need for the institution to hold any information about me other than that I exist and that I am subject to its regulations for such and such a time period; there is no need for me to remember anything about my own past, or any sets of skills beyond a fairly simple motor set. Why I am there and who I am just don't matter to the institution itself; it "remembers" all it needs to know through the complex set of procedures that it puts into place. Sima Qian, a Chinese historian (ca. 145–86 BCE), made a similar observation about the burning of the books in 213 BCE. Qian (1994) writes:

> Approving his proposals, the First Emperor collected up and got rid of the *Songs*, the *Documents*, and the saying of the hundred schools in order to make the people stupid and ensure that in all under Heaven there should be no rejection of the present by using the past. The clarification of laws and regulations and the settling of statutes and ordinances all started with the First Emperor. He standardized documents. (31) [The translator notes that this refers to the standardization of bureaucratic practices, not of the script.]

This replacement of memory by procedures extends to a formal information processing argument that Ross Ashby made about closed systems all kinds. He argued that if we completely know a system in the present, and we know its rules of change (how a given input leads to a given output), then we don't need to bring to mind anything about the past. Memory, he said, is a metaphor needed by a "handicapped" observer who cannot see a complete system, and "the appeal to memory is a substitute for his inability to observe" (Ashby 1956, 115). Now no institution is ever total, nor is any system totally closed. However, it remains true that there are modes of remembering that have very little to do with consciousness. These modes tend to abstract away individuality (extension of a person back in time) by substituting rules and constraints on the behavior of types of people for active recall.

At the other end of the spectrum is the hypermemory of Funes, the Memorious, discussed by Jorge Luis Borges. As the result of a riding accident, Funes had a perfect memory; however, it was so good that it took him far longer to recall an instant than it had to experience it:

> Funes remembered not only every leaf on every tree of every wood, but every one of the times he had perceived or imagined it. He determined to reduce all of his experience to some seventy thousand recollections, which he would later define numerically. Two considerations dissuaded him: the thought that the task was interminable and the thought that it was useless. He knew that at the hour of his death he would scarcely have finished classifying even all the memories of his childhood. (Borges 1998, 135–136)

Funes's memory repeats the past; it is sequential; in this way it is like the memory of Luria's mnemonist. He has no random access to different parts of his youth, though he tries to create the same by classifying and enumerating his experiences. This is a fractal memory—each act of remembering calls up worlds of experience, and each world calls up new acts as complex as the first. Operating socially, this non-discriminatory memory is often a political tool; Baudrillard (1995) writes that the deluge of information about the Gulf War was "to produce consensus by flat encephalogram" (68). At this end of the spectrum, we get total individuality: there is no such thing as a generalized bin in which to store selected past data, with the only possible redemptive act being classification.

Across this span from no active recall to hyperawareness there is a dazzling array of memory practices that we engage in on a daily basis; there are few censuses of these practices. What is really interesting is not so much the individual practices and how they articulate a given set of memory technologies. Rather, it is how sets of memory practices get articulated into memory regimes, which articulate technologies and practices into relatively historically constant sets of memory practices that permit both the creation of a continuous, useful past and the transmission sub rosa of information, stories, and practices from our wild, discontinuous, ever-changing past.

This possible object of interest (and obscure object of my desire), memory practices, extends in space into a unit I will call the archive and into time in units I will call the epochs of memory.

The Catalog of Traces: The Archive

Just What Is the Archive? The Landlord tells us that time is the money of science. It is indeed one of the coins. To carry on philosophical and scientific commerce, we have historically needed to agree on units of measurement of time, space, and process. Sometimes this takes the form of agreeing to very precise units of measurement. Mr Swivett, for example, is complaining about the establishment of the Gregorian calendar in the United Kingdom. To talk to each other (or a forteriori to work in parallel), two computers generally need to share a common clock; my own gets its time from the atomic clock server in Denver, Colorado, so that it can communicate effectively with some boxes in Santa Clara, California. Rather less precisely, geologists need to fix their epochs in order to be able to translate results from one corner of the earth to the other. Further, they have had, historically, to negotiate the kinds of packages that time comes bundled in: is time basically a formless line or does it have a shape, so that our planet was once young and thrusting but is now middle-aged and flat? When I swap stories with my colleagues at the university, I know that there are various well-accepted patterns to time; a current obsession among many of my kind is the ever-receding (never proven) golden age when universities were universities and there was no need to be constantly on the make as one produced theories. Although these seem like heterogeneous examples, I do not see in principle much difference between them: in order to carry out effective communication, we need to be able to share units and shapes of time.

The Landlord points to scientists being interested in events that occur at different points of the earth at the same instant. What he does not say—it is so obvious perhaps—is that a structure of record keeping will subtend this common time, rendering it useful through permitting the collocation of accounts of said events. Scientists make (im)mutable mobiles (Latour 1987). Let us refer to this structure of record keeping as the archive. "Arkhe," Derrida (1996) notes, "names at once the *commencement* and the *commandment*. This name apparently coordinates two principles in one: the principle according to nature or history, *there* where things *commence*—physical, historical or ontological principle—but also the principle according to the law, *there* where men and gods *command*, *there* where authority, social order are exercised, *in this place* from which order is given" (1). He names these two orders sequential and jussive; and he asserts that from this point on (from the inception of the *arkhe*), "a series of cleavages will incessantly divide every atom of our lexicon" (ibid.).

We will come back to beginnings at numerous points in this book, but let me point out at the start that in the beginning is the inaugural act: the moment from which memory is assumed to be perfect and time to begin. In his exploration of the history of writing, Clanchy (1979) tells us quite bluntly that the 'fixed limit' of the validity of written agreements of September 3, 1189: "which continued for the rest of the Middle Ages, marked the formal beginning of the era of artificial memory" (123). He has told a complex story in three times—first, it came to be recognized in England that written documents could be trusted as well as printed ones; second, it came to recognized that one had not only to generate documents but also to store them in an archive; and third, it was seen that techniques of reference access to the library were needed in order to render them useful. For there was, he points out, no great reason to think that the written record should be preferred over the memory of trusted witnesses. What we would call forgeries could be conceived of as documents written to justify the ways of God to men (ibid., 148). Patrick Geary, telling a similar story over a different time period in France, argues that what happens with the development of a written memory was the supplanting of the role of women as memory keepers by men in monasteries. He argues for a similar inaugural act—the keeping of written memories was part of a move to consolidate new power relations by partly by creating false continuities and discontinuities with the past: "Arnold of St. Emmeram compared the process of sorting through the past to the process of clearing the arable, cutting down groves once sacred to the gods so that the land could be made useful for the present. This same pruning was going on in archives across the continent. Both he and Paul of St. Père de Chartres emphasized that not everything was to be preserved, only that which was useful" (Geary 1994, 114). Geary points out that future generations of historians have been held to the documents produced by this inaugural act of winnowing and fashioning, and so have tended to see the first millennium as a more radical break with the past than it probably was.

Each major change of storage medium over the past several centuries has engendered proclamation of similar inaugural acts. As we will see in chapter 1, Charles Babbage (1837) proclaimed that until the invention of printing, "the mass of mankind were in many respects almost the creatures of instinct" (59). Now, the great were encouraged to write, knowing that "they may accelerate the approaching dawn of that day which shall pour a flood of light over the darkened intellects of their thankless countrymen," seeking "that higher homage, alike independent of space and time, which their memory shall for ever receive from the good and the gifted of all countries and all ages" (ibid., 54) . For him, this marked the true commencement of our species. There are more than enough kinds of time here to keep the Landlord happy. There is the inauguration, which we today would put at 1453, of the era of intelligence for the mass of mankind. There is acceleration (things moving fast) and timelessness (homage being outside of space and time). In print mediated communication, this latter timelessness has often been seen as a central feature—marked, for example, by Landor's imaginary conversations (1824), which juxtaposed quotes from the great and wordy in such a way as to form fluent conversations across time and space.

We are perhaps not quite at the point of witnessing the inaugural act for the archive of computer-mediated communication, but its prophets are many. One relatively sober form comes from Avi Silberschatz and Jeff Ullman (1994): "There is now effectively one worldwide telephone system and one worldwide computer network. Visionaries in the field of computer networks speak of a single world-wide file system. Likewise, we should now begin to contemplate the existence of a single, worldwide database system from which users can obtain information on any topic covered by data made available by purveyors, and on which business can be transacted in a uniform way" (929). Computer scientists have frequently announced the dawning of a new age. In chapter 2, I explore Auger's claim (1960) that "now, after the age of materials and stuff, after the age of energy, we have begun to live the age of form" (ii). The old age, he argued, was one of diachrony and materialism: it gave us the historicist visions of Darwin and Marx. This age, he argued, is that of synchrony and form. When such an epistemic break is operated, the knowledge of the previous age becomes irrelevant; when the break is constituted by the move from diachrony to synchrony, the past is doubly deleted. There are many analogous inaugural acts for perfect memory systems woven into the fabric of our history. Lavoisier's chemistry textbook inaugurates the modern era of chemistry by forging discontinuities with past chemistry (changing the names of substances to remove relationships with alchemy; not mentioning continuities with previous work (Bensaude-Vincent 1989)). Lyell's *Principles of Geology* (1830–1833) does much the same—attaching a catastrophic time (schools of thought erupting onto the landscape but then going nowhere) to prior geology and a uniformitarian time to his own. The rhetoric goes that there is nothing worth remembering from chemistry or geology beyond these inaugural acts; but that after these acts each chemical or geological contribution will be remembered time out of mind.

For Derrida, the archive is not only sequential back to an origin, it is also jussive. It tells us what we can and cannot say: "The archive is first the law of what can be said, the system that governs the appearance of statements as unique events" (Foucault 1982, 129). My reading of these claims is not particularly Derridean or Foucaultian. The jussive nature of the archive comes down to the question of what can and cannot be remembered. The archive, by remembering all and only a certain set of facts/discoveries/observations, consistently and actively engages in the forgetting of other sets. This exclusionary principle is, I argue, the source of the archive's jussive power.

Three examples indicate the nature of my claim. In an article called "Setting Limits to Culture," Ian Hunter argues that the academic field of cultural studies has tended to fall into an aesthico-ethical reading of culture, even when it was avowedly materialist. He notes that administrative change of the type carried out by Kay-Shuttleworth in the mid-nineteenth century (he was a leading advocate of universal education) tends to get written out of the cultural histories—even though his work had a lot more to do with the founding of the state, say, than the arguments held by political economists. Hunter (1988) asks: "Why then are we predisposed to ascribe thinkers like Engels and his more famous partner—or, for that matter, prophets of culture like William Morris or Matthew Arnold—central roles in the process of cultural development, and to consign administrative intellectuals like Kay-Shuttleworth to the relative obscurity of educational history?" (105). His response is that on the whole, academic attempts to look at the forging of organizations and the framing of cultural attributes are carried out "in the shadow of a single general process of contradiction, mediation and overcoming at whose end lies the 'fully developed' human being" (ibid., 106). Putting this in a completely different way, the memory of infrastructural change is not held overtly—if it is held at all, it is held in the most abstract forms furthest away from it (in the form of a memory of intellectual manifestoes epiphenomenal to the infrastructural change).

Mary Douglas describes the consistent institutional forgetting by the discipline of psychology of a number of independent discoveries of the social or collective nature of memory. Following Donald Campbell, she asserts that "it is professionally impossible in psychology to establish the notion that institutional constraints can be beneficial to the individual. The notion can be scouted, but it cannot enter the memorable corpus of facts" (Douglas 1986, 83). She goes on to note that, ironically but naturally, Campbell forgot his own insight and turned to biological determinants. Douglas claims that this eminent forgettability is due to the discovery not fitting in with the institutional commitment to individualistic methodologies—in other words, there was no place for the facts to be pigeonholed.

Finally, Yrjo Engestrom (1990) points to the difficulty that ethnography has in examining the concept of memory: since in general ethnographies deal in very thin time slices, but memory is accreted over months, years, generations; equally, ecological studies have often been limited by the career span of the ecologist, who finds it difficult to further a career with a one-hundred-year experiment, say. Not so much "man" as the measure of all things but our careers. The set of stories that we can tell about the past strings together facts and fancies that we can justify collecting in the present. The gaps are wide—as we will see in chapter 3, the set of stories we can tell about life and its history are massively weighted by the difficulty of getting grants to study parasites and viruses. Indeed, one of the chief problems in relating political economy to scientific thought is that we hold our knowledge of the two in such impermeable containers—both in terms of data bins and people.

Hunter and Douglas point to a somewhat idealized feature of the jussive nature of the archive: the fact that what ought to be remembered is all and only that which fits in with the worldview legitimated by the inaugural act. Typically, Engestrom is somewhat more mundane; he tells us how this forgetting or overlooking can take place in practice. He notes that the archive contains a set of methodological rules for the accretion of facts and theories, and that certain kinds of facts and theories just cannot fit. The edict (thou shalt not write about social memory) is translated into a fact about the world (there is no way in which such and such a kind of data can be gathered). The archive's jussive force, then, operates through being invisibly exclusionary. The invisibility is an important feature here: the archive presents itself as being the set of all possible statements, rather than the law of what can be said.

But as I write this, I am aware of the hypostatization that is going on here. There is of course no single archive; we as a society operate multiple sets, far more heterogeneous than functionalists like Douglas could ever see from behind the walls of their archive. The motivation for the singular designation of the archive is twofold. I do want to talk about it in the singular, first, because I am trying to describe features common over the set of archives that we construct. And I do believe (though this remains to be shown) that there are sets of dependencies between archives that lead to regularities among the exclusions (and commonalities among the inaugural acts). This is a degree zero of the archive. Patrick Tort's study (1989) of the rise of genetic classification systems in the nineteenth century demonstrates how there has been a filiation between archival principals operating across a wide range of fields; he traces links between the fields of the classification of writing, linguistic typologies, race classifications, and criminal physiognomy, for example. A second motivation is that this locution points to the fact that memory can be highly diffuse and so it can be useful to think in terms of a generalized archive. It is only in pathological cases that the memory of how to perform a given organizational task is fully held in one person's head, in one computer or filing system, or even fully within that organization. Organizations delegate memory tasks to the environment. I delegate to my tax accountant the memory of my previous years' tax records; to my employer the tasks of remembering exactly what my income has been and of sending me an appropriate form; to my filing system a set of possible deductions; to my head a rough idea of how much I can afford to claim on various dubious items (working out what my notional budget with the Internal Revenue Service is); to a tax guide for college professors a list of esoteric deductions that I should consider. This highly diffuse, technologically mediated memory is the stuff of memory practices.

The Act of Remembering So one would imagine that one of the first things to do in this singular archive is to stock things in memory. Indeed, the fact that they are being stored in memory is absolutely no indication that they will ever be recalled from memory. Clanchy, we have seen, made this point about medieval archival practices. However we operate much the same relative autonomy between the act of remembering and the act of recall ourselves in daily life. Thus Serge Tisseron (1996) has written eloquently about the act of taking a photograph. We often see this as a memory act. Looking at it this way alone, Tisseron writes, does no justice to the countless thousands of undeveloped films there are in the world, or the similarly numberless set of developed photos that get jammed into a drawer or tossed into a box without ever being looked at again. Rather than see this as a failed memory act, he suggests that we devote more attention to what happens when we record something for the archive. What happens, he says, is that we frame the present moment through the act of consigning it to another medium for storage. We compose a picture that expresses something about the way we are. The act of taking a photograph is an act of conceptualizing the present: this is an important moment; this is how I see my brother; this is my friend. The act of remembering frames the present in a particular way: it is a tool with which to think.

Much archival practice involves an isolated act of recording. I share the academic passion for photocopying and filing away articles, which I have no real intention of ever reading. I filter listserv discussions into huge email files that I imagine I will get to some day, but whose usefulness times out well before I ever do. I take notes at meetings and at lectures, knowing full well that I'll probably just throw them away afterward. Am I an archival misfit, a broken record keeper? I think not (though I'll have to check my notes). This litany can be repeated again and again for collective memory practices—though it rarely finds itself in the official literature of any particular archive. John Gillis (1994) writes that we have as a society reached a frenetic pitch with our multiplied acts of remembering: "Every attic is an archive, every living room a museum. Never before has so much been recorded, collected; and never before has remembering been so compulsive" (14).

Doctors incessantly take medical histories when they greet patients, refusing to accept previous histories from other doctors or from themselves: the act of taking the history does a great deal of communicative work in the present; the notes are frequently incidental. Generations of data have been lost with changes of information technology. The John Rylands library at the

University of Manchester in England houses, for example, a collection from the Jodrell Bank radio telescope of basically unreadable printouts of data from early computers. We collect vast amounts of biodiversity information about the planet (terabytes per day; from satellites, ground observations, aerial photography) that are dutifully archived but never actually analyzed—there are not enough people on earth or techniques to look at it all. Al Gore memorably referred to this as data rotting in an information silo. These collective cases, as in the case of the individual photographer, constitute a way of framing the present. They often indicate a drive to render the world memorable and thus governable (Foucault 1991)—a way of acting on the present rather than recalling the past. The League of Nations deliberately archived inaccurate data in the 1920s while in the process of setting up a procedure for keeping mortality and morbidity data across the world. The early record keeping constituted a form of disciplining doctors and citizens to the need for this data to be systematically collected. The act of rendering memorable does not mean that at any stage it will be remembered.

The Scope of the Archive In a working archive, facts are among the last things that are actually stored—both for individual archives in the form of memories stored to meat and for social archives in the form of file folders, journals of record, and so forth.

In a notable experiment, psychologist Edouard Clarapède had a "masked and disguised individual" break into his lecture theater, spout some nonsense words, wave his arms about, and then head for the door. Two weeks later, students had a very fuzzy recall of what had transpired; Clarapède inferred that the spectacular and the isolated cannot easily be stored in memory—we are best at stocking the routine and the mundane. He concluded that the past "even of a simple event—is less a record than a sort of taxonomy. Not perceptions, but categorization of familiar types was the major function of the memory. Our testimony depends much less on our memory, than on the mental image that we possess of a type or a class in which we arrange facts" (Matsuda 1996, 109; cf. pp. 95–98 on Bergson).

Similarly, Daniel Schachter (1996, 103–104) describes experiments to determine recall of items on a list—a generic category like "sweet" might not be on the list containing "chocolate, sugar, good, taste, tooth and bitter," but it will be remembered as being a member. The only people who score well on excluding it are amnesiacs!

Transitioning to the public sphere, the classificatory dimension of memory can be seen clearly in the memory theater of Guido Camillo described by Frances Yates (1966). The mnemonic device that Camillo used is the partitioning of all possible events into the rows and tiers of a notional theater, with mythological, Christian, and astrological registers vying together in a synthetic classificatory rage. The effect is a verbal version of a yantra, wherein architectural, Buddhist, and mythical registers imploded to hold massive amounts of knowledge in a single painting.

James Fentress and Chris Wickham (1992), in a work reminiscent of Frances Yates's, argue that artificial memory systems waned after Descartes: "Instead of a search for the perfectly proportioned image containing the 'soul' of the knowledge to be remembered, the emphasis was on the discovery of the right logical category. The memory of this system of logical categories and scientific causes would exempt the individual from the necessity of remembering everything in detail. . . . The problem of memorizing the world, characteristic of the sixteenth century, evolved into the problem of classifying it scientifically" (13).

A. J. Cain (1958) demonstrates that Linnaeus's binomial classification system took the form it did (specifically the number of genera; Linnaeus argued that botanists should be able to recall the names of all genera) in order to be easily memorable. He did not believe that botanists could remember all species names. However, some of the earliest incunabula were field guides for botanists—and one can speculate that this classificatory turn in the seventeenth and eighteenth centuries was directly linked to the expansion of artificial memory systems with the rise of the book.

One of the main jobs that paper archives do is to consolidate a classification system that makes it possible to forget the particular. Auguste Comte ([1830–1845] 1975) speaks of this directly in his *Cours de philosophie positive*, where he states that in this current positive age, we no longer need to remember precisely the turns in the trail that led to a specific scientific discovery—any facts can be unambiguously classified into his unchanging, complete schema, and so the personal details can drop away in any account one might wish to concoct. Indeed, he argues, they *must* fall away for else there is too much drain on our memory faculties: "The constant tendency of the human spirit, as regards the exposition of knowledge, is therefore to progressively substitute the dogmatic order, which alone can suity the perfected state of our intelligence, for the historical order. . . . It would be certainly impossible to reach the desired end, if one wanted to make each individual spirit submit to passing successively through the intermediate stages which the collective genius of the human species necessarily passed through (ibid., 65). We are not in general able to remember complete stories about the past. There is an overwhelming amount of evidence both individual and social that this is not what we do well. What we do well is to disaggregate a fact about the past into a number of standard elements, and then set in train a procedure for reassembling the specific out of the general. This sets in motion a system of memory recall that is able at any given moment to create a working version of the past.

What is stored in the archive is not facts, but disaggregated classifications that can at will be reassembled to take the form of facts about the world. (We will look later at processes of commemoration and memorialization, which are special cases that do seek to save the particular.)

The Place of the Archive In a wonderful passage in his *Principles of Geology*, Charles Lyell discusses the earth as an archive commissioner. He was working from the position that there was no sign of the origin of the earth, or any portent of its end: what we have access to is a set of records in the landscape that leave the impression of massive upheaval and discontinuity in the past. This was strongly at odds with his picture of the earth as being subject now to the same forces as ever—with the appearance of massive change being wrought by a vast increase in the amount of time afforded the geologist to account for the face of the earth. The gap between appearance and reality was the record-keeping process:

> Let the mortality of the population of a large country represent the successive extinction of species, and the births of new individuals the introduction of new species. While these fluctuations are gradually taking place everywhere, suppose commissioners to be appointed to visit each province of the country in succession, taking an exact account of the number, names, and individual peculiarities of all the inhabitants, and leaving in each district a register containing a record of this information. If, after the completion of one census, another is immediately made after the same plan, and then another, there will, at last, be a series of statistical documents in each province. . . . the commissioners are supposed to visit the different provinces in rotation, whereas the commemorating process by which organic remains become fossilized, although they are always shifting from one area to another, are yet very irregular in their movements [so that] . . . the want of continuity in the series may become indefinitely great. (Lyell 1830–1833, 3:31–32)

This passage prefigures a major theme of this book: the tools that we have to think about the past with are the tools of our own archive—so that we generally project onto nature our modes of organizing our own affairs (just as we tend to understand the brain in terms of the dominant infrastructural technology of the day—from nineteenth-century hydraulics in Freud to the telephone switchboard in the 1920s to network infrastructure today). However, we will not dwell on that for the time being. Rather, let us look at what this text says about record keeping.

First and foremost, Lyell is saying that the earth itself is a sort of record keeper—perhaps not a very good one, but a record keeper nonetheless. Geologists today have expanded this record-keeping function enormously, seeing traces of the distance past (beyond revolutions in the earth's surface and even before the creation of the earth) in various isotope ratios. Our earth weaves its own history into its texture. Similarly, life itself writes its history into the earth. The very oxygen that we breathe has been freed through the metabolic processes of cyano-bacteria to enter the atmosphere. Massive hard sea floors have been created by the disaggregated exoskeletons of planktons; these floors have fostered the development of new forms of life. Without life, the earth would be an inhospitable place for life: with the positive feedback loop in place, it has become more and more livable for an increasingly complex set of organisms. Inversely, without the earth as it is, life might well be simpler; the current relative peak of biodiversity (abstracting away anthropogenic extinction) is sometimes argued to be a feature of the more complex geology of the shattered supercontinent Pangea (Huggett 1997, 299). The lesson here is that, with the introduction of life, the traces that we leave of the past are neither other from us nor passive: they render life more livable.

This brings us to the parable of the ants on the beach, adumbrated by Simon and commented on by Ed Hutchins. Simon's original story runs that if we look at ants moving on a beach, we might impute their complex trajectories to internal programming, rather than being a fact about the beach. Hutchins (1995) invites us to go outside of the normal time constraints of the psychologist (shades of the exclusionary principle) and look at the beach over a several-month period:

> Generations of ants comb the beach. They leave behind them short-lived chemical trails, and where they go they inadvertently move grains of sand as they pass. Over months, paths to likely food sources develop as they are visited again and again by ants following first the short-lived chemical trails of their fellows and later the longer-lived roads produced by a history of heavy ant traffic. After months of watching, we decide to follow a particular ant on an outing. We may be impressed by how cleverly it visits every high-likelihood food location. This ant seems to work so much more efficiently than did its ancestors of weeks ago. Is this a smart ant? Is it perhaps smarter than its ancestors? No, it is just the same dumb sort of ant, reacting to its environment in the same ways its ancestors did. But the environment is not the same. (169)

This seems to me a good reading; it evokes the generalization that one of the things that all life does is to transform its environment by leaving memory traces in it, thereby increasing its chances for success. Further, Hutchins suggests, a snapshot view of this complex will have a given organism reacting relatively intelligently against a passive backdrop, whereas the complex {environment + entity}, mediated by archival practice, is in fact the seat of intelligence. Bruno Latour (1996) has playfully suggested that a difference between people and animals is that the former accrete memories from the past in technology: this is how he distinguishes between the perpetual reassertion of rank among baboons and the more placid acceptance of a given social order among humans. To the contrary, it is a characteristic of life itself to leverage its work practices through engineering its archive.

Let us move from the ridiculous to the mundane. Maurice Halbwachs (1968, 52) put it beautifully: "most groups . . . engrave their form in some way upon the soil and retrieve their collective remembrances within the spatial framework thus defined" (Halbwachs 1968). James Walsh and Gerardo Ungson (1991, 65–66) spoke of "ecology," or physical design, being one of five "storage bins" for organizational memory—the other four being individuals, culture, transformations (procedures), and structures (roles). A few examples of such engraving will help. Imagine you are alone in a forest—take the forêt de Fontainebleau outside of Paris. You want to go on a walk through the wilderness. You could get a map, and rely on the paper archive to provide your trail through the area. On the other hand, you might equally well choose a color (yellow for the easy walks, and black for the difficult ones, where you have to scramble) and follow the ribbons attached to trees and the streaks painted on rocks full circle. And if a few of the ribbons or streaks are missing, you just follow the track of the footprints (you would be wise to do so where they indicate that the majority of the people have gone off the trail for a distance, possibly to avoid an obstacle). Simon Schama (1995, 546–560) writes at length about this reworking of the Fontainebleau wilds as the first set of guided walks in Europe, though of course from Hansel and Gretel on into historical time we have reworked our natural landscape to leave a memory trace. We often don't think of such trails in memory terms, because it is not our own personal memory that is being engraved—it is the collective memory of our culture. We operate such changes a fortiori in the built environment. If you visit a Catholic church, you don't have to remember the order of the Stations of the Cross. They are laid out for you in a standard fashion. Our reorderings can be evanescent. I constantly litter my morning path with objects that I want to remember to take to work—books to go to work in a pile next to the bathroom; clothes for dry cleaning on the hood of my car; things I really must do today on my computer keyboard. The generic trick I am using here is putting matter out of place as a form of aide-mémoire. (I have given here a trio of somewhat functional examples: the memory that we hold in the built environment is by no means necessarily useful memory—as indicated by the well-known saga of the carriages on Roman roads making ruts of a certain size that propagate through to the nineteenth century along with axle design, since a nonstandard axle will always be straddling ruts. Then onto train tracks, since that is the axle size that smiths were used to building, and so forth).

On this wide reading of the archive, I am folding distributed, diffuse archival practice into the sets of specialized archival technologies: the list, the file folder, the computer database, and so forth. These latter constitute one small subset of an extremely large set of memory practices that we engage in from day to day, from century to century.

Why All This Folding? So I have partially removed the act of remembering from the telos of recall and folded it into a reading of ways of being in the present; I have partially removed the operation of the archive from the telos of the storage of useful facts (the scope of the archive); and I have partially removed archival practice from any set of specialized storage technologies (the place of the archive). In each case, I have folded the archive into our sets of actions in the present and in the built and shaped environment. Taken together, these three work to decenter analysis of the archive from the rather terse *Oxford English Dictionary* definition: "A place in which public records or other important historic documents are kept." The central reason for this decentering has been to suggest that if we want to look at memory practices, we should not look at cases of the good or bad recall of facts stored in specialized media (brains, files, or disks). Recall is frequently irrelevant; when it is, the facts of the matter are frequently irrelevant; and even when we have recall of facts as the goal, specialized media are frequently beside the point. If we want to understand memory practices in the sciences or in other spheres, I am suggesting, then we need to look elsewhere. My goal in this book is to begin to trace some delineations of this elsewhere.

It is clear from this folding that a lot of our memory practices are about the present: we should not be looking for them in dusty archives. Consider the first. This generalizes to the observation that classification systems come into being at the site of a memory trace (a play on Freud's remark, developed by Benjamin, that "consciousness comes into being at the site of a memory trace" and thus "becoming conscious and leaving behind a memory trace are processes incompatible with each other within one and the same system" (Benjamin and Arendt 1986)). One way to put it would be to say that we classify in order to be able to forget: we impose a classification in the present at the site of a fact that we would otherwise have to remember. With the grand imperial bureaucracies that developed in the nineteenth century, innumerable ways were developed of classifying people into sets so that rather than having to "remember" that such and such a person was, say, a criminal, one could reconstruct the fact at will from their classificatory pigeonhole, as offered by criminal physiognomy or phrenology, for example (Tort 1989).

Patrick Geddes, writing at the height of Victorian certainty, provides a typical expression of this imperial drive. He broke the world of knowledge down into Sciences (not historical or statistical), with the sole member being Logic; Sciences (statistical but not historical), namely, Chemistry, Physics, and Mathematics; and Historical and Statistical, which includes Ethics, Sociology, Psychology, Biology, Geology, and Astronomy. The territory of a given society—a fraction of the scheme—would be broken down as follows:

A. Territory of given society
Quantity at given time.
Persistent since last unit time.
Added since last unit time
By geologic agency (upheaval, deposition, etc)
By social agency (discovery, conquest, reclamation, purchase, etc.).

II. Quality at given time.
Unused.
Used.
Unspecialized (for such and such functions).
Specialized (for such and such functions).

Decrease since last unit time.
By geologic agency
By social agency. (Geddes 1880–1882, 305)

Many such classification systems were produced during the nineteenth century. Geddes gives the following justification:

The scheme is scientific throughout—in accordance with the known truths of physical and biological science—is capable on the one hand of complete specialization by the aid of minor tables, into the most trivial details of common life, and on the other, or generalization into a colossal balance-sheet. Its systematic and generalized character appears clearly from a survey of the whole sheet of tables. It will be observed in the first place that the successive sets of tables, three each, may be read in horizontal rows, thus—Territory, Production, Organisms, Occupations, Partition, User, Result. Secondly, that these sets of tables are related to each other: Organisms being treated on the same plan as Territory; the tables of Occupations being derived largely from those of Production, and the tables of Partition, User, and Result, being in such close relation to those of Occupations that the ruling of each of the latter is exactly copied in all the four lower series; while the third, and by far the most important general view is obtained by looking at the left hand and middle vertical series (at least as far down as Occupations inclusive, and in some respects all the way), as entries on the debtor side of the balance-sheet, and similarly at the right hand vertical series as entries on the creditor side. Again, the scheme is universal in application—the tables will serve equally well for arranging our knowledge concerning any society—animal or human, civilized or savage: for savage and animal societies, some columns here and there of course simply remaining blank. (ibid., 310–311)

The move here is double. First, we will create a huge balance sheet, wherein outputs and inputs can be fully squared away, between the natural and social worlds and within both. Second, with the balance sheet in place, we will be able to read off laws about nature and about societies—the more complete the classification scheme, the less we will need to know about a particular case, the less we will need to hold in memory.[4] (The balance sheet was also the trick that allowed Lyell to get beyond the historicism of his uniformitarian vision into a set of general laws.)

Whereas the first folding encourages us ultimately to look to the relationship among classification, forgetting, and memory practices; the second encourages us to look at standardization, forgetting and memory practices. I introduced it earlier in terms of our shaping of the physical environment. Two central facts about such shaping—be it done by bacteria, ants, or people—are that it is socially negotiated and that it comes in standard packages. We can only draw off the wealth of archival practices we inscribe into the landscape efficiently to the extent that we share a set of conventions about their meaning—it is for this reason that Leroi-Gourhan (1965, 64) referred to the natural unit for memory as not the brain but the ethnic collective.

A significant part of the setup work for networked information infrastructures is putting into place a set of agreements, which should be remembered from that moment on—the jussive aspect of Derrida's archive. As the Landlord said: "The Philosophers need a Time, common to all, as Traders do a common Coinage" (Pynchon 1997, 192). Ontologies flow from these agreements. Thus, for example, there really is no single model of just how the Internet works and what should be done at the physical level and what should be left to programs. In an early edition of Tanenbaum's *Computer Networks*, a typical passage reads: "Although Section 1–3 is called 'Network Software' it is worth pointing out that the lower layers of a protocol hierarchy are frequently implemented in hardware or firmware. Nevertheless, complex protocol algorithms are involved, even if they are embedded (in whole or in part) in hardware" (1996, 20). "Protocols" is a key word here. The earliest use of the word, according to the *Oxford English Dictionary*, is in 1541 and its central

4. And this classification can have formative dimension. See the classifications of science by Comte and Ampere and the science of Zetetics developed by Tykociner in the twentieth century: a hole in the balance sheet means knowledge is missing.

meaning is as follows: "The original note or minute of a transaction, negotiation, agreement or the like, drawn up by a recognized public official, notary, etc. and duly attested, which forms the legal authority for any subsequent deed, agreement, or the like based on it." Internet protocols carry this meaning; they are records of a transaction between negotiating bodies at the same time as they are formal procedures for communication between parts of a network. Any given standard sits along a continuum. It can be inscribed into the world in such a way that one's affordances for action are constrained to its observation; it can operate as a verbal commitment; or it can be "hardwired" into institutional or informational technologies. But whatever the form, it needs to be shared.

Standards are not only jussive—by virtue of their inaugural act, they are also sequential. When a radically new standard is introduced, there is a new starting point for history; from now on, everyone who adopts the new standard will be compatible with each other adopter. Within the information world, there is generally some attempt to make new standards backward-compatible (so that, say, your new version of Microsoft Word can read documents created by previous versions). However, the general point remains that the new standard word processor marks a point around which new applications can be developed, new training courses be designed, and so forth. The power of thinking about standards as memory practices like any other lies not so much in what it tells us about standards per se, but in what it tells us about memory. It is clearly only through a very artificial set of divides that we can separate analytically the domains of classification and standardization from the field of memory practices. If we accept the divides, then we cannot hope to understand memory practices of our species—how we configure the world and ourselves to maintain an active memory of the past. If we reject them, then we can start to significantly widen our understanding of the nature and workings of memory. As a general rule, memory operates in the present and socially through a variety of "dispositifs techniques" whose principal tools are classification and standardization. The recording of individual facts about the past in an individual brain is not foreign to this process; it is, however, but a small part. Equally, the recording of facts garnered in laboratory experiments constitutes but a tiny fraction of memory practices in the sciences.

Which brings us back to the frenzy of memory practices that motivated the first folding. We should not think of the archive as an organized set of data to which we desire random access. This puts too much emphasis on the noun, on the site. Rather, one of our chief ways of dealing with the world is to remember things. The act of remembering sometimes coalesces into more or less complete classifications, more or less rigorous standards. But more generally, it is one of our chief ways of being in the world as effective creatures: it is a way of framing the present; a mode of acting. We exude, sketch, form, design memory traces of all kinds; these work together in complex ecologies. Our task is not to track the site of memory (to the Ark), but to describe the circulation of memories through the multifarious media we have developed.

The Epochs of Memory

In 1832 Eugène Sue published a novel about Ahasuerus, the wandering Jew whose refusal to offer succor to Jesus on the road to his crucifixion translated into the doom of walking the face of the earth, unable to stop, unable to die. Ahasuerus had taken it upon himself to save his family: he tried to arrange for a huge fortune to go to his final descendants in industrializing Paris (Sue 1844). Unfortunately, the Jesuits had known about this money for some two hundred years, and they had been tracking these same descendants, trying to either ensnare them in the church or otherwise neutralize their presence. Three kinds of memory are pitted against one another in the novel: the recollection by Ahasuerus of his own past; the voracious record keeping of the Jesuits, described in loving detail; and Christ's memory of the original crime (perhaps held by St. Peter the record keeper and finally resolved in the death of his descendants). Sue's story weaves its way through these three memory registers, exploring in beautiful detail the features and fragrances of each.

Ahasuerus's goal is sweet oblivion—a rest from his tireless, endless state of consciousness and the memories that assail him. When justice has been done—when his line becomes extinct through the cruelty of the Jesuits—then Christ releases him (betimes) from his curse. The theme is a common one; we have seen in the West the development of a trope that when justice has been done the aborigines in Australia or the Native Americans and African Americans in the United States, through an act of apology, then the past can be laid to rest. Past iniquities will be forgotten by most people and institutions: when justice has been done in the present, then their memorialization will be complete (and they can be pushed out of consciousness). Is it possible, Yosef Yerushalmi (1996) asks, referring to the trial in France of Klaus Barbie for war crimes, "that the antonym of 'forgetting' is not 'remembering,' but justice?" (117).

At issue here is the periodization of memory. The Christian church has created a very long present, in which we are held responsible for the sins of Adam and Eve: the inaugural act of eating from the tree of knowledge has affected all subsequent generations. Other practices are more forgiving, though the line of demarcation is generally hard to draw. In the case of a nation, what needs to be remembered can stretch for hundreds of years (e.g., in Ireland, where the politics of the sixteenth century live today) or can be legislated to be as short as the previous week. David Cressey (1994) writes that "during the 1670s there was a memorial calendar for the Tories and a competing regime for the Whigs, each with its high cultural and popular dimensions. At the same time there were counter-memories, suppressed memories, even legislated Acts of Oblivion, to extinguish the deeds of the revolution" (69). And certainly nowadays most English hold no memory of their revolution. Matt Matsuda (1996) cites French historian Renan's observation that "forgetting, and I would even say historical error, are essential factors in the creation of a nation; in this, the progress of historical studies is often a danger for *nationality*. Historical investigation, in effect, brings to light facts of violence which took place at the origin of all political formations, even those whose consequences were the most beneficial. Unity is always brutally created" (206). Scientific disciplines range from considering any paper that has already been published as significantly out of date (high-energy physics, where preprints are the coin of the natural philosopher) to requiring the month and day of publications from the eighteenth century in order to determine naming priority (systematics).

We have seen earlier that archives begin with inaugural acts that wipe out a past and define a period "from now on" as the present. It is possible to distinguish a set of memory epochs emerging from new media of record keeping. Jacques LeGoff (1992), following André Leroi-Gourhan, refers to five epochs: "The history of collective memory can be divided into five periods: oral transmission, written transmission with tables or indices, simple file cards, mechanical writing, and electronic sequencing" (54). The boundaries between the epochs are rarely clear—Homer's epics were consigned to writing and then print, Ugaritic texts have made it to the World Wide Web, and the canon of classical Greek literature is on CD-ROM (in the form of the *Thesaurus Linguae Graecae*). These remain real boundaries, however, in two ways. First, every act of migration across media is a conscious act in the present: unless there is a contemporary constituency for a book, for example, it will not find its way onto the Web. Most never make it across—just as, in small, most government archives never make it across a new filing system. Second, the migration itself changes the document: a given rendering of *The Odyssey* could only become definitive with the new medium of print—it was highly "fluid"; a sprawling nineteenth-century novel can be read in new ways in electronic form, where we can create automatic concordances and character searches on the fly.

Each new medium imprints its own special flavor to the memories of that epoch. In *L'homme nu,* Lévi-Strauss (1971) contrasts the time of myths and the time of novels, attaching the former to the oral medium and the latter to the printed word. In a myth, he argues, episodes can occur in almost any order in different versions of the same story; it is only with the novel that we get the fixed sequencing of linear time. Eisenstein (1979) goes so far as to forge a causal link between our conception of linear universal time (time moving onward along a continuum at the rate of one per second) and the rise of the printing press—with the sequential ordering of facts on a page itself opening our imagination to a new conception of time. The causal argument has to be refined, for Sohn-Rethel's attachment (1975) of absolute time to the creation of the commodity form is much more compelling; however, there certainly is an apposition between the printing press and the efflorescence of a new kind of temporal framework within which information could be readily stored and easily accessed. And it is certainly true that such a new temporal (or spatial) framework for documents of itself bears a significant degree of the ideological load. Thus Arthur Miller (1991), writing about sixteenth- and seventeenth-century temporal and spatial frameworks in Oaxac, Mexico, notes: "The colonial calendars inscribed a new political and economic significance on the Zapotec conception of temporality. Incorporation of European writing into the form of the ancient calendar perpetuated the concealing of sacred knowledge, while at the same time it undermined the ancient structures of authority. Similarly, the native map tradition was a response to the European secularization of land ownership, while it also tended to destroy land ownership's ancient sacred connotations" (143). Equally, there is an apposition between the new media for communicating goods and information of the nineteenth century—the railway and the telegraph—and the imposition of standard time reckoning across Britain and America (Chandler 1977): "philosophers," as the Landlord said, needed to be able to create a unified archive for a vast expanse of territory. With this new time, events can be sorted into archives operating with new temporal and spatial units: they can be processed very differently.

A further periodization of memory and memory practices can be generated from changing attitudes to the archive. It is generally accepted (though the details still require better elucidation) that there was a rise of historical consciousness in industrializing Europe in the early nineteenth century. By the middle of the century, Nietzsche (1957) was referring to this as a crisis: "I am trying to represent something of which the age is rightly proud—its historical culture—as a fault and a defect in our time, believing as I do that we are all suffering from a malignant historical fever and should at least recognize the fact" (4). Ever the iconoclast, Nietzsche is one of a small group of writers who have written paeans to the virtues of forgetting[5] (Yerushalmi 1988, 9).

5. We return to this theme later.

Two somewhat independent sources select the period 1870–1914 as being particularly significant in the recent history of memory practices. The first is constituted by historians such as Eric Hobsbawm and David Cannadine. Hobsbawm (1992) argues: "Once we are aware how commonly traditions are invented, it can easily be discovered that one period which saw them spring up with particular assiduity was in the thirty or forty years before the first world war. One hesitates to say *with greater assiduity* than at other times, since there is no way of making realistic quantitative comparisons. Nevertheless, the creation of traditions was enthusiastically practiced in numerous countries and for various purposes" (263).

Cannadine (1992), in a study of British monarchy and the invention of tradition between 1820 and 1977, points to "an efflorescence of 'invented' ritual and tradition in Wilhelmine Germany and the French Third Republic" (103) and to a renewed ceremonial and putatively traditional configuration of the British monarchy in the period between 1870 and 1914, a period when "in London, as in other great cities, monumental, commemorative statues proliferated" (128). This was also the period of the invention of the Mafia out of a mixture of whole cloth and puppet theater, and of its projection onto a distant past (Fentress and Wickham 1992, 173–199). This period can be characterized, then, as one of fervid creation in Western Europe of a particularistic past (marked by commemorative plaques, statues, rituals). It is also, argues Ian Hacking (1995), the period of the situation of "memoro-politics" at the heart of intellectual discourse: "Today, when we wish to have a moral dispute about spiritual matters, we democratically abjure subjective facts. We move to objective facts, science. The science is memory, a science crafted in my chosen span of time, 1874–1886" (220). Pointing to the political role of memory, Hacking questions its scientific certainty. Two figures who are greatly admired by Hacking are Freud and Proust. LeGoff (1992) accuses Freudian psychoanalysis of being part of "a vast anti-historical movement which tends to deny the importance of the past/present relation" (16)—this around the issue of Freud's concentration on perceptions of the past over what actually happened. Hacking makes a similar charge. Over this period, Proust's stunning, particularist memory continually replays the history of the aristocracy (from the name of the country to the name of the people); his Charlus both stands most on the ceremony of this history and is most vulnerable to an uncovering of his own personal, masochistic story (Proust 1989); Charlus chooses an aristocratic tradition to hide what he is doing in the present. The new memoro-politics uncovered by Hacking in the science of psychology takes the same form as the manipulation of tradition discussed by Hobsbawm.

This apposition of Hobsbawm and Hacking prompts a highly tentative generalization. There was, perhaps, something going on toward the turn of the twentieth century concerning the deliberate texturing of the past in order to work in the present. True representations of the past are irrelevant for both streams—the invented tradition is about the assertion of a current emergent political reality; for Freud, especially, the past may be lost but it needs to be reworked in order for the client to be able to live in the present. There are, of course, many invented traditions that one can point to, spanning the centuries. If the marking of the period from the 1870s holds, it is as one where memoropolitics took central stage in artistic and social scientific representations of the individual and in representations of the state.

It is, integrally, a period when massive new waves of information classification and standardization took place—international classifications were developed for diseases, work, criminal physiognomy, and so forth: facts could be split apart, sorted into pigeonholes, and reassembled in new ways. It is a direct outgrowth of this work at the turn of the twentieth century that we get the emergence of the database as a central cultural form. Lev Manovich (2001) puts it beautifully when he writes: "As a cultural form, database represents the world as a list of items and it refuses to order this list. In contrast, a narrative creates a cause-and-effect trajectory of seemingly unordered items (events). Therefore, database and narrative are natural enemies. Competing for the same territory of human culture, each claims an exclusive right to make meaning out of the world" (225). Manovich develops a syntagm/paradigm couple, where the syntagm represents a statement that is made, and the paradigm represents the set of possible statements. He argues that with the new technology, "database (the paradigm) is given material existence, while narrative (the syntagm) is dematerialized. Paradigm is privileged, syntagm is downplayed. Paradigm is real, syntagm is virtual" (ibid., 231). The observation obtains, but its inception should not be attached to the new computing technology. Rather, the current status of databases completes the movement begun in the nineteenth century of universalizing classification systems.

One can see Manovich's argument becoming true in fine in the development of database technology this century. The first commercially available computer databases were organized hierarchically. If you wanted to get to a particular piece of information, then you went to the overarching category and made a series of choices as this category broke down into groups, then subgroups, until you got to the specific piece of information that you required. This mode of traveling through a database was called "navigation." The next generation, network databases, followed the same logic. The user had to follow one of a number of predefined pathways in order to get to the data—it was more ordered than a straight narrative archive but still pre-imposed a set of narrative structures on the data. The following generation, relational databases, began to break this mold. The underlying database structure is a set of relations or tables, each table having rows and columns. This matrix form allowed a new form of enquiry to be made: you no longer had to travel the preset pathways, you just had to declare what you wanted to know in a controlled language. Finally, object-oriented databases operate on the principle that you don't need to know either pathways or relationships beforehand: each data "object" carries its salient history with it, and pathways and relationships can be in principle reconfigured at will (Khoshafian 1993, 114–121). Frances Yates begins her *Art of Memory* (1966) with a contrast between the great geniuses of two different ages—Aquinas and Einstein. Aquinas was recognized a genius, she claims, because of his prodigious memory; Einstein because of his brilliant thinking. We can add a third term to this sequence, with the development of the human genome database. The canonical scientific act for our times (sequencing the genome) resonates with the social and technical turn to non-narrative memory described by Manovich.

I will give a name to the current epoch, the site of the memory practices explored in this book, by calling it the epoch of potential memory. To continue Manovich's trope, this is an epoch in which narrative remembering is typically a post hoc reconstruction from an ordered, classified set of facts that have been scattered over multiple physical data collections. The question is not what the state "knows" about a particular individual, say, but what it can know *should the need ever arise*. A good citizen of the modern state is a citizen who can be well counted—along numerous dimensions, on demand. We live in a regime of countability with a particular spirit of quantification. Foucault (1991) pointed out that this is one of the principles of governmentality: a modern state needs to conjure its citizens into such a form that they can be enumerated. The state may then decide what kind of public health measures to take, where to provide schooling, what kind of political representation should be afforded, and so forth. Uncountables in the West are our version of the untouchables in India: a caste that can never aspire to social wealth and worth. In order to be fully countable and thus remembered by the state, a person needs first to fit into well-defined classification systems. At the start of this epoch, the state would typically, where deemed necessary, gain information on its citizens through networks of spies and informers writing narrative reports; such information gathering continues but is swamped by the effort to pull people apart along multiple dimensions and reconfigure the information at will.

But that seems to be quite a jump, from the way in which databases work to the operation of the state. The jump is possible because our way of organizing information inside a machine is typically a meditation on and development of the way we organize it in the world. When the first object-oriented language, Simula, was invented, it was perceived as a way of modeling the way that things were actually done in the world. The claim today is still that you take a simple English-language description of system requirements and turn the nouns into objects and the verbs into operations, and you are up and running. Object oriented programming, by this claim, is the transparent language bar none. At the same time, and from the other end, numerous management theorists claim that now that we have object oriented programming, we can reconfigure the organization so that it matches the natural purity and form of the programming language. No longer do we need hierarchical modes of communication; we can organize according to teams with their own sets of interfaces with management, but where management does not need to know how any particular job is carried out by the team. Thus a programming language that operates as part of an organizational infrastructure can have potentially large effects on the nature of the organization, through the medium of organization theory. So object orientation is on the one hand a model of the world, and on the other hand the world is learning how to model itself according to object orientation. This kind of bootstrapping process is common when you deal with infrastructures. Generally, I would describe it as the programming language and organization theory converging on a particular instantiation of the organization in which object-oriented programming will furnish the natural, transparent language. This convergence is central to information infrastructures. We make an analytical error when we say that there is programming on the one side with its internal history, and organization theory on the other with its own dynamic. The programming language is very much part of the organizational history and vice versa. James Beniger (1986) made this kind of connection in his work. Following a robust tradition in cybernetics, he noted that in the late nineteenth century many things came together to make process control a key factor in management and technology.

Ours is certainly not the first society to hold memory primarily in nonsequential form—indeed, this is precisely the point that Lévi-Strauss made about myth; or that one could make about the memory devices of the Luba (Roberts et al. 1996) or about Tibetan yantras. However, I would argue that it is this turn, begun in the nineteenth century in the office and in government agencies, that takes us out of the age of the book. JoAnne Yates (1989) traces this transition beautifully in her work on late-nineteenth-century office technology. The earliest correspondence books, she notes, held painstaking transcriptions (or, later, blotted copies) of outgoing letters in chronological sequence. The two great revolutions in office technology, she noted, were the manila folder and the hanging file drawer—these together permitted the rearranging of data into subject files. Later copying technology (notably the invention of carbon paper) allowed a single piece of information to be stored in multiple places. As this technological work was going on, she notes, there was also a withering away of the greetings and salutations in internal correspondence—so that the new genre of the office memorandum was created, which in turn gave rise to the genre of email. And at the same time, information that previously had been collected in narrative form (if at all) was now distributed into statistical tables (Chandler 1977).

We have seen, then, two characteristics of the current memory epoch: greatly increased centrality of the reworking of the past for the operation of the state (Hobsbawm and Hacking) and greatly increased technical facilities for such reworking with the development of database technology. We are getting to be very good at reconfiguring the past as a tool for exploring/supporting the present. The past that we are colonizing in order to do this work is not "wie es eigentlich gewesen ist." To the contrary, the canonical archival forms of the present tell the past as it should have been. Comte, I think, sets the tone for this whole period with his assertion that we cannot afford to keep in our own minds (and to pour into the minds of children) what actually happened in the history of science. There is now, he said, too much science out there for this to be feasible. Rather, what we should do is classify the sciences completely and tell stories about each science that show the logical steps that brought us to our current state of knowledge—a move that nowadays we would call "rational reconstruction." When the new political tradition (the changing of the guards; the Mafia) is created, it tells the story of a past that should have been, in order for current political conditions to be justified.

We return here to the aporia spelled out previously, but this time with an understanding that it is characteristic of memory practices in the current epoch across a wide range of activities: from science to politics to business.

Conclusion

Discussing the daily life of the Greek gods, Sissa and Détienne (1989) write that the mythographers conjured a divine existence in which "time does not pass, it is frozen and collected into an eternal present." For, they argue, any external influence from events, people and so forth would pose the gods as incomplete. And yet, they note, poets such as Hesiod had no difficulty imagining densely packed works and days for the gods. One of the founding myths of scientific practice is that science is carried out in an eternal present, from which all external influence has been banished. For, it is said, if a cat walks by a machine designed to detect gravity waves, then the measurement is invalid (Collins 1985). And yet this present is an outcome of densely packed daily practice by scientists—a practice that has its chroniclers in the sociology of science.

The time that is the coin of the philosopher is, in Tanaka's words (2004), a time "synchronized" over many disciplines and practices. It is the outcome of a massive work of building organizations, classifying the world and its inhabitants, and integrating material from multiple domains. The resulting eternal present and linear chronology are imperfect products. The time of mitochondrial DNA has at times conflicted with that of evolutionary theory, the time of physics has clashed with the time of geology (Burchfield 1990). However, in a messy, sprawling, gargantuan sort of a way it is a towering achievement. An achievement that beetles over its own particular mnemonic deep. The memory practices that are the subject of this book require an analysis of the textures of this time and of that deep. These practices constitute a space out of which the former and the latter are generated in all their generality and idiosyncrasy.

This book constitutes an attempt to chart out that space. Chapter 1 deals with the nineteenth century and with synchronization. It demonstrates the mapping of both the social and the natural world into a single time package: located in the eternal present of the proximate future (social time) and deep reality (natural time). Through the temporal mapping, mediated by the metaphor and the materiality of information technology, a pure second nature was created that was fully archivable despite apparent discontinuity. Chapter 2 explores cybernetics in the mid-twentieth century. I uncover a different mapping of nature and second nature through the evacuation of memory—there is no particular past, only kinds of past. I argue that both cybernetics in the 1950s and geology in the 1830s packaged time in ways indexed by their information technology so as to permit a traffic between the social and the natural worlds. I foreshadow the argument that attributes these disparate disciplines and epochs share—an eternal present, an evacuated past, synchronization through temporal packaging—are features of the longue durée of knowledge in the West over the past few hundred years. Chapter 3 celebrates and excoriates the database as a central symbolic artifact of the past few centuries. I endeavor to show how the very material substrate of the archive can play a central role in our understanding of the past. Chapter 4 evokes the spaces, entities, and times that are excluded from the synchronization effort in the sciences of biodiversity today. The basic argument there is that this mnemonic deep is a textured mnemonic deep; the hope is (following Michel Serres) that my contribution will be to open up other possible spaces and times for human enquiry. Chapter 5 asserts, as baldly as may be, that the stories we (the globalizing "we") tell about the past through our dizzying array of scientific practices are simultaneously a representation of our political economy through the prism of our information technology and a denial of that representation in an attempt to universalize as we globalize. A concluding chapter consigns this book to oblivion.

1

Synchronization and Synchrony in the Archive: Geology and the 1830s

"LECTURE FIRST. Chronometricals and Horologicals, (Being not so much the Portal, as part of the temporary Scaffold to the Portal of this new Philosophy)."

—Plotinus Plinlimmon in Herman Melville, *Pierre or The Ambiguities: The Kraken Edition*

"Bacon's brains were mere watch-maker's brains; but Christ was a chronometer; and the most exquisitely adjusted and exact one, and the least affected by all terrestrial jarrings, of any that have ever come to us."

—Plotinus Plinlimmon in Herman Melville, *Pierre or The Ambiguities: The Kraken Edition*

Le monothéisme impose et suppose un constant travail de la mémoire.

—François Hartog, *Le miroir d'Hâerodote: Essai sur la repràesentation de l'autre*

The mathematician and the natural philosopher had assumed to themselves the highest locality in the temple of science, and had almost expelled the collector and the classifier from its precincts. Presuming that magnitude and distance ennobled material objects, and invested with sublimity the laws by which they are governed; and taking it for granted that the imponderable and invisible agencies of nature presented finer subjects of research than the grosser objects which we can taste, touch, and accumulate, they have long looked down upon the humble and pious naturalist as but a degree superior to the functionary of a bear garden, or the master of ceremonies to a cage of tigers. This intolerable vanity—this insensibility to the unity and grandeur of nature, to the matchless structure of sublunary bodies; and to the beautiful laws of organic life, was perhaps both the effect and the cause of the low state of natural science during the preceding two centuries. Men of acute and exuberant genius were naturally led to invest their intellectual capital in researches that were likely to return them an uxorious interest in reputation; and it must be acknowledged that the richest fields of science were for a long time left to the cultivation of very humble laborers.

—Anonymous, "Life and Works of Baron Cuvier"

Epigraphs: Herman Melville, *Pierre or The Ambiguities: The Kraken Edition* (New York: HarperCollins, [1852] 1995), p. 301; Herman Melville, *Pierre or The Ambiguities: The Kraken Edition* (New York: HarperCollins, [1852] 1995), p. 303; François Hartog, *Le miroir d'Hâerodote: Essai sur la repràesentation de l'autre* (Paris: Gallimard, 1991), p. 13; Anonymous, "Life and Works of Baron Cuvier," *Edinburgh Review* 62 (1836): 277.

Introduction

Information technology constitutes the twist in the Möbius strip that takes us from arguments internal to a field (how is the past conceptualized in the case of a historical science like geology) to its exterior (how is information about the past stored). Generally speaking, all things on earth can be seen as at once objects and archives (Buckland 1997). As objects they function in the world, and as archives they maintain traces of their own past. Thus a rock can be read as an object that constitutes part of the lithosphere, and equally as a document that contains its own history written into it: striations on the surface indicate past glaciation, strata indicate complex stories of deposition over time, and the relative presence of radioactive isotopes of various kinds indicates, among other things, journeys through the mantle. And so forth. The significance of these traces has been discussed in a variety of contexts within the sciences over the past few hundred years: the span is from Paley's natural theology (1802) (timeless divinity) to Allègre's new historicism (1992) (perpetual change). In general, we have gone from a very limited view of what we can know about the past to a much more ambitious view. In the early nineteenth century, Charles Lyell argued that we could not know any more about the history of the earth than the period since the last revolution of its surface; today methods have been developed to trace material—with varying certainty—through subsumption into the earth's mantel, and even back to its interstellar origin. (Similarly, Comte in the early nineteenth century argued that we can only ever know anything beyond mere physical dynamics about those bodies in space that we can touch—he saw the solar system as our ultimate limit; today we claim, improbably, to reach back in space toward the origin of time itself (Weinberg 1993).)

As we have come to see the earth as a variably legible palimpsest from which little is ever completely lost, archival technology has itself exploded over the past few centuries in two ways in the earth sciences—from underneath, with the use of technological practices to visualize and to read archives contained in the earth, and from on top, with the use of computer technologies to manage the associated information explosion. (I am here adopting a soft definition of technology—drawing off Foucault's "dispositif technique" (1975), which refers to integrally social and technical sets of practices that constitute particular actions in the world. The soft definition is needed in order to be sensitive to the fluid nature of both vanilla and information technology. Office procedures and computer programs are both designed as abstract machines and so can be realized interchangeably—though always with great difficulty—in configurations of people, the environment, and silicon.) The Möbius strip that concerns us now, then, is one that goes from reading the earth's archive to archiving the earth in the 1830s. The archival metaphors used converged onto information practices on the one hand and earth science on the other to create a closed system where each bootstrapped the other.

Three themes run through this chapter. The first is that of synchrony and stasis. In a historical science like geology produced at the height of the Industrial Revolution in England, one might expect a progressivist telling of the history of the earth. To the contrary, we will wander a strangely synchronic world—where there is no effective arrow to time, where there is in the long run no change, and where humans rather than transforming the world with the might of industry are really having no effect on it at all. The second is that of the two natures—industrial and natural—converging onto this isotropic time. The third is that of synchronization—the work that it takes to bring the various bits of the world together into a single archival framework. Information practice—its metaphors and strategies—permitted the cohabitation of astronomy, political economy, industry, and geological science in this synchronic world, and, in particular, it permitted a synchronization of the social and natural worlds to the same temporality.

In and of itself, this claim of the importance of information practices is not proven or disproven by the work of this chapter alone—there are many possible readings of the material presented. Any proof will cumulate over the set of readings offered in the next three chapters—discussing geology in the 1830s, cybernetics in the 1950s, and biodiversity today. At each point, a central science for an epoch will be shown to be associated with an information discourse structured around the latest information technology and speaking to the histories associated with these disciplines. This line of argument in the end posits radical discontinuities in the disciplinary record of sciences in tune with the development of such discontinuities in information technologies broadly conceived (Veyne 1971, 395–424). Why these discontinuities, which have the form of Foucault's epistemic breaks, should occur as they do is discussed in the conclusion.

Time and Memory in an Industrial World

Time management was a central issue in industrializing Britain in the early nineteenth century. Charles Babbage (1832) considered the clock a "regulator of time" in opposition to "the negligence and idleness of human agents" and "the irregular and fluctuating effort of animals or natural forces" (32–43). A century before the glorious rise of scientific management, Claude-Lucien Bergery (1829) remarked: "The worker . . . must be mean with his time . . . he can hardly devote 30 years or 262,800 hours to collecting the money he will need in his old age (180) . . . Each minute lost will deprive him of about threethousandths of a franc (181) . . . every man is capable of at least 5 movements a second, there are 36,000 seconds in a day of ten hours, which will in consequence allow 180,000 movements" (108). The working day in a factory was regulated by clocks, from the checking-in clocks that often registered the hours of the workers to the timers that made the machines run regularly. Dickens at the start of the nineteenth century and Lewis Carroll at its apogee both give accounts of the significance of timekeeping—to young boys (David Copperfield) and white rabbits—that would have appeared startling to any eighteenth-century social commentator. When the great geologist Charles Lyell, whose work I examine at length in this chapter, underlined the importance of the division of the history of the earth into equal periods of time, he was reflecting a fundamental obsession of the industrial world then being born.

Varieties of technoscientific work developed during the nineteenth century created a representational space and time that structured both the immediate perception of social and natural reality and the storage of these perceptions in an imperial archive (Richards 1996). The imperial archive is often, and rightly, pictured in terms of its voracious desire to collect information—of all kinds, about all of its citizens all of the time. It can be seen as a lust for information that is reaching a kind of apogee today. However, there is another way of picturing this information drive. It is also a drive to save as little information as possible about something; it relentlessly pares down interactions to their absolutely minimal form and then dispenses dialogue by rote. An anecdote—I have to call it this since I cannot find the trace—might make this clear. When trams were introduced into Melbourne, so the story goes, the tickets were designed to contain information about the passengers. You didn't want someone (say a tall, thin person) giving away her ticket to someone (say, a short, robust one). So the conductor would punch holes in the ticket to indicate the physical appearance of the passenger along several dimensions. In a stagecoach, in days of yore, every passenger would be known by the driver—a lot of information would be held about the passenger. With the early Melbourne tram, the amount of information decreased. Finally, with the tram today—when we are duly disciplined to the logic and the logic has been duly adjusted to us—we have tickets with two-hour time limits that can be used multiply over their period of validity. On today's reading, the cost of interaction (constantly checking tickets, verifying identities) is too high, so only the trace of a valid transaction carried out by someone within the recent past is really all that's needed. The empire voraciously gathers as little information as it can. In

Latour's terms, the oligopticon is the flip side of the panopticon. This imperial archive cannot be understood solely in terms of record keeping. The act of keeping records is itself intimately tied to the conjuring of the social and natural world into forms that render themselves amenable to recording. Thus the act of archiving is a positive act that simultaneously changes the world and creates a record of it (cf. Serres 2001, on smallpox)—in Derrida's terms, it is both sequential and jussive.

I am not concerned here with the origin of the modern structuring of the scientific and social archive. The scientific use of isotropic, coordinate space and time clearly predates the period I am looking at. However, it is to the nineteenth century that we must look for the growing belief that scientific record keeping—and, in particular, its way of representing nature and society within that space and time—could be applied to all possible social and natural phenomena. In the 1830s, Comte's Church of Positive Philosophy stands as an extreme statement of the scope of science and of the need to radically reshape the archive; by the end of the century, such scientific fervor and archival manipulation was commonplace (Rosenberg 1976). This neutral space and time coordinate frame became plausible for two tightly interlinked reasons. First, a series of infrastructural technologies were developed that actively extended the applicability of the framework. Second, the bureaucratic working procedures developed in association with these technologies took advantage of and formalized this framing. Scientific work epitomized the development and rigorous application of these procedures; here my work resonates with Serres's theory of the origins of Euclidean geometry in social science and the administration of the Greek empire (1993).

The Industrial Revolution and Regular Space/Time

One of the chief early products of the Industrial Revolution in Britain was the watch: from 1800 to 1820 at least 100,000 clocks and watches were produced every year in England (British Sessional Papers 1818, 205)—at a key period between 1831 and 1841, the census revealed a jump from 8,000 to 15,000 people employed in this trade. Literary texts of all sorts from this period teem with arguments from analogy to watches. Thus animals could be understood through this analogy: "As a watch not wound up remains without motion, still retaining the power of resuming it, and when the mainspring recovers its elasticity is again enabled to act upon its wheels: so to animals *heat* is the key that winds up the wheels, and restores to the mainspring its powers of reaction" (Buckland 1836, 159). The analogy extended to the physical features of humanity, as evidenced by the following appeal by Neil Arnot (1827) for "medical men" to learn physics: "All these structures the medical man, of course, should understand, as a watchmaker knows the part of the machine about which he is employed. The latter, unless he can discover where a pin is loose, or a wheel injured, or a particle of dust adhering, or oil wanting etc., would ill succeed in repairing an injury, and so also of the ignorant medical man in respect to the human body" (xxviii). It also extended to the turbulent ocean. As D. Graham Burnett (2003, 5–19) notes, Matthew Fontaine Maury spoke of a "clockwork ocean" with waves and cycles of salinity as "balance wheels."

The analogy permitted a comparison between two orders of creation through imposing a temporally mediated qualitative difference between creator and created—a difference that was instantiated on the factory floor and in nature. Creators stood outside regular space and time and imposed regular temporal order. In general, scientific work was seen as the imposition of a representative framework of regular space and time on social and natural time. This is brought out in a work by Ajasson de Grandsagne and Parisot, French translators of Lyell's *Principles of Geology* (itself a work whose vision of a changeless world did much to develop regularities in space and time as a key to the representation of global process). They started from the position that "industry, so it is said, is a second nature" (Grandsagne and Parisot 1836, 8). A historical section demonstrated how humanity had gone from being initially weak to currently king of nature. Science, they argued, caused the change, and was now all-powerful: "Thus, in raising itself above everything else, science has achieved its full extent: all the arts have been submitted to it, industry has recognized it for its regulator; it has served and protected humanity in all its states, and it has interwoven itself in the most intimate and sensitive manner with all social relations" (ibid., 21). Science had emerged from previous intellectual disorder to govern this second nature: if, in principle, science has had some element of chance, and if common people have made some useful advances, henceforth it is only through the meditation of superior spirits that it can spread new benefits (ibid., 20). Thus science stood in the same relation to the second nature—industry—as did God to the first; the new scientist created order in artifacts as God did in the motions of the material world. Science was industry's "regulator": a word that refers to the timing of the industrial process. Science was making society conform to the same spatiotemporal representational framework it was "discovering" in nature: a process that I call "convergence" (Bowker 1994). This is a key aspect of the synchronization of the social and natural worlds.

Neither nature nor second nature were, of course, as regular as this spatiotemporal framework: and indeed there developed a strain of politically radical science that attacked Newtonian physics for its attempted imposition of such a form of representation (e.g., Mackintosh 1840). Thus ex-St. Simonian Jules Leroux wrote of a different kind of time in science: "Beside the philosophy of human history is placed the philosophy of natural history; for there is in fact only one science. *The cosmogony of nature and humanity.* To make science, art and politics converge more and more towards the same goal, to introduce more and more in science, as in art, as in politics, the notion of change, progress, succession, continuity, life, and by that to submit them to the same law: this is the goal, the outline, the plan of philosophy" (*Réformateur*, October 16, 1834, 4). Radical geologist Constant-Prévost Daurio (1838) put ethereal motion at the center of his system and excoriated the space of "that dry and scowling science, which has for head a sphere powdered with *alphas* and *betas*, a cube for body, cylinders and cones for arms and legs; which lives only in square and cube roots, and doesn't offer the world anything but formulas, which dandy geometers find very elegant; a science which can only move with the aid of winches, pulleys, and the steam engine. Lacking sensation, it is sensitive as a monolith" (220).

Conservative scientists and political philosophers endeavored to demonstrate that beneath the appearance of chaos and irregularity—both in first and second nature—could be found the reality of regularity. This time was quintessentially spacelike: it lacked direction and granularity. At first blush, this might seem an odd representational framework to develop for the brash progressivism of the Industrial Revolution (compare the progressivist historical vision of the science of Victorian certainty, as described in Young 1973). However, as acute observors such as Dickens and Zola noted across the span of the century, the primary apperception of industrialization may well have been the irregularities of boom and bust as much as the annihilation of space and time (I am thinking here of David Copperfield's and Denise's tortuous working childhoods—marked more by staccato syncopation than regular rhythms). Tracing the logical structure of this archival framework in two domains—political economy and astronomy—I will begin to demonstrate its attachment to synchrony and synchronization.

G. Poulett Scrope was a geologist and political economist. As a geologist, Scrope (1833) was very sensitive to the importance of time: "The leading idea which is present in all our researches, and which accompanies every fresh observation, the sound of which to the ear of the student of Nature seems continually echoed from every part of her works, is—Time! Time! Time!" (165). When it came to political economy, the trouble was that people were not regular: "The rules of Political Economy are as simple and harmonious as the laws which regulate the natural world, but the strange and wayward policy of man would render them intricate and difficult" (ibid., title page). From a similar vantage, Charles Babbage (1832, 3–49) asserted that the machine regularizes "man": processes synchronizing him to industrial time . . . which is also natural time. For Scrope, it is the capitalist order. Thus he wrote of the periodic slumps that were much more visible to the early Victorian observer than the unremitting progress that was to be the mark of later Victorian science and political economy: "It is . . . strongly to be suspected that such epochs of general embarrassment and distress among the productive classes . . . are anomalies, not in the order of events which flow from the simple and natural laws of production, but occasioned by the force of some artificial disturbing cause or other, introduced through the fraud or folly of the rulers of the social communities they so grievously affect" (Scrope 1833, 152). Here we see already an astronomical influence: astronomy at this period was "saving" Laplacean determinism through the analysis of anomalies. Scrope makes the analogy tighter still: "There is, in fact, a continual oscillation . . . going on in the returns of capital in most employments, about the mean level or average of *net profit*." This, he claimed, went along with or was, rather, caused by "an analogous oscillation in the Market value or *selling price*, of commodities about the *mean cost of their production*" (ibid., 162–163).

The same horror of waywardness—and its solution through the imposition of a regular spatiotemporal representational framework—was common in astronomy in this period. Consider the following proud announcement concerning Halley's comet: "The return of the comet of Halley at its predicted time has been remarked with intense curiosity and satisfaction by astronomers, and by the public. This has now become a regular and well ordered member of our system" (Anonymous 1836a, 160). The author claimed that this was a good augur for the development of science: "That a body, differing so much in its appearance and habits from the ordinary tenants of the sky, should reappear so nearly at the time and in the place appointed for it by our excellent associate Rosenberger, is a convincing instance of the progress towards certainty of physical astronomy" (ibid.). The regular flow of time, itself inspired by the watch analogy, was the central discovery of astronomy, and thence of the human spirit: "In effect, we only know completely one single law; that is the law of constancy and uniformity. It is to this simple idea that we seek to reduce all others, and it is uniquely in this reduction that science, for us, consists. . . . Such is, I believe, the natural movement of the human spirit . . . of which Astronomy offers us the clearest image" (Poinsot 1837, 386–387). Geology's task was to bring this representational framework down to earth. The result is typified by Huot (1837, 3): "Everywhere one sees such a uniform disposition, that only differs in a few details . . . that often presents lacunas but never inversion. This order that one admires, despite so many traces of violent revolutions, of upheavals and shocks that the Earth has felt, does it not seem, if one dare say it, in rapport with the regular march imprinted on celestial bodies?" (3). This, as we will see, was Charles Lyell's whole project.

To complete the circuit back from the heavens to humanity, historian and geologist Philippe-Joseph-Benjamin Buchez (1833) believed that the greatest good would emerge from imposing this astronomical representational framework (based on infinite space and a timeless present) on our understanding of human affairs. Thus he wrote that, to all appearances, "harmony is nowhere, not even in the circle of the smallest families" (6). This apparent disharmony would, with the proper analytic framework, be dissolved into regularity: "When one examines the position of humanity vis-à-vis the phenomenal totality in which it exists, one easily gets to see that it is a function of the universe, in the full mathematical rigor of this term. . . . Thus one comes to understand that the very large revolutions of humanity correspond to small revolutions of the planetary system (ibid., 109–110). In a similar move, political theorist Charles Dunoyer (1837) averred that the underlying astronomical order would, through the "spirit of industry," assert itself on humanity:

> Under its [the spirits of industry's] influence, peoples will begin by grouping themselves more naturally. . . . Peoples will come closer together, mass according to their real analogies and according to their real interests. Given this, the same arts will soon be cultivated with an equal success among all peoples, the same ideas will circulate in all countries . . . ; even languages will get closer . . . ; uniformity of costumes will be established in all climates no matter what the conditions of nature: the same needs, a similar civilization, will develop everywhere. . . . and finally the largest countries will end up by only representing a single people, composed of an infinite number of uniform aggregations, between which will be established, without confusion or violence, relations both as complex and as easy, as peaceful and as profitable as may be. (I: 452–453)

In this Laplacean era, then, both astronomical and political bodies would converge onto regular time and henceforward be fully deterministic.

In sum, a cosmology inspired by factory production developed in the early nineteenth century in Britain and France. According to this cosmology, humanity was working on two fronts. First, disordered human events were being made orderly through the development of the capitalist mode of production on the one hand and by the coupling of wayward people with regular machines on the other. Second, the human spirit was discovering that all apparent disorder and disharmony in the wider universe (as revealed to astronomers and geologists) could, with the appropriate representational framework, be dissolved into a regular ordering of processes in periodic time.

In each case there was an appearance of disorder and secular—or progressive, taken with the emphasis on "stepping" rather than "forward"—change (the appearance of comets, booms, and slumps); in each case, the underlying reality was an eternal present discovered through the use of a representational framework of temporal regularity. The convergence of these two fronts meant that each could find rhetorical and philosophical support in the other.

Within a fully regularized temporal framework, one can read backward into the past and forward into the future, without recording the shocks of the present. The real archive told a tale of stasis. The emphasis is on a single Law, with perturbations and oscillations providing noise that true understanding would filter out. There is no need (as we shall see in Comte's analysis in chapter 2) to actually record the past as it happened; all one need record in order to act now is the past as it would have happened if there had not been perturbation. The nature of the archive itself changes (over time, with the perfection of the human spirit) from being one of the chronicling of difference to a synchronized display of sameness; within which there is no falling back or leaping forward but rather the smooth operation of regular law.

An ideologically charged eternal present lurks beneath the veneer of change and underwrites the archive. Such lurking is native to significant political and scientific argumentation today. Consider the case of climate change. The political framework for the debate is that time should stand still (the current climate is perfect and natural; a hard argument for Mancunians, whose city grew during the industrial revolution partly because of the desire of textiles for moisture rather than humans for comfort, to follow), and in particular human agency should be removed from the flow of time and rendered nugatory. In such a perfect world, we would be subject to regular cycles (orbital forcing of climate change for example), which would—in the long run—be as nothing. One can structure a reading of biodiversity politics very similarly: the goal of much current policy is not the preservation of change and its potential but the preservation of current species and the imposition of stasis. (It can of course be argued that what is being opposed in climate change arguments is the catastrophic rate of change rather than the fact of same—in the same way that it was feared universally in the 1970s that the catastrophic onset of the next ice age was nigh, unless perhaps the U.S.S.R. could seed the air with carbon dioxide (Lamb 1995, 333). It is my reading that in general arguments about rates of change are really about the fact of change). The act of record keeping would become one of mapping the apparently wayward present onto an eternal, unchanging present ever just out of reach. Beyond dharma, at the heart of the industrial revolution, there is stasis.

Synchronization of the World and Its Archive into Cartesian Space/Time

Shortly after the development of the locomotive, Wolfgang Schivelbusch (1986, 18) asserts: "the uniform speed of the motion generated by the steam engine no longer seems unnatural when compared to the motion generated by animal power; rather, the reverse becomes the case" (18). He opines that mechanical uniformity became "the 'natural' state of affairs, compared to which the 'nature' of draft animals appears as dangerous and chaotic" (ibid.). This latter was a favorite theme in industrializing Britain and France in the nineteenth century.

Indeed, the abstract space and time within which scientific representations have been made since Newton are in turn reinforced by the railway. Schivelbusch (1986) speaks of the train "realizing Newton's mechanics in the realm of transportation" (59). He makes the rich point that what was left to the train traveler (after you took out sounds, smells, and so forth from the countryside) was a Newtonian world of abstract qualities—size, shape, quantity, and motion. He goes on to point out that there was a definite learning process involved. Early railroad travelers literally could not look out of the window at all: they found it disturbing, confusing; it gave them headaches. Only when they learned to deal with the new form of representation on its own terms were they able to look out the train window and appreciate the landscapes that lay before them.

This new space created a new kind of geographical representation: abstract and regular. Schivelbusch speaks of the "annihilation of space,", by which he means the de facto and representational annihilation of lived space between. William Cronon (1991) has described this same effect in a passage describing the long-distance transportation of meat in refrigerated railway cars: "Once within the corporate system, places lost their particularity and became functional abstractions on organizational charts. Geography no longer mattered very much except as a problem in management: time had conspired with capital to annihilate space. The cattle might still graze amid forgotten buffalo wallows in central Montana, and the hogs might still devour their feedlot corn in Iowa, but from the corporate point of view they could just as well have been anywhere else" (259). It is perhaps unsurprising that it was within the featureless great plains of the Midwest that space should first be annihilated!

Historians of America's Great West have been struck by the commodification of nature that occurred in the midcentury. This commodification itself was tied to the development of new means of communication (especially the railway and the telegraph) and the associated new ways of doing business. Across the vast, flat prairies, Cronon (1991) notes, the land came to resemble the maps that were drawn of it by government surveyors in the distant east. Fields, fences, and firebreaks "were concrete embodiments of the environmental partitioning that made farming possible" (102). The Midwest was indeed peculiarly suited to the government practice of "subdividing the nation into a vast grid of square-mile sections whose purpose was to turn land into real estate . . . . By imposing the same abstract and homogeneous grid pattern on all land . . . government surveyors made it marketable" (102). As Cronon says, the grid: "turned the prairie into a commodity" (103); he talks of a "second nature" being created by the railways, one that came to seem quite natural (56). Thus again representation in an abstract, regular space and commodification go hand in hand. Globalization is an ever-incomplete movement to impose a uniform representational time and space on a heterogeneous collection of lived spaces and histories.

The story of the imposition of a new social and representational time has been well told by Schivelbusch, Chandler, and Cronon. As with the annihilated space, this story is clearly tied to organizational needs. Schivelbusch (1986) gives the sparsest account: "Regular traffic needs standardized time—this is quite analogous to the way in which the machine ensemble constituted by rail and carriage undermined individual traffic and brought about the transportation monopoly" (48) (for a fuller account, see O'Malley 1996, 55–98). The analogy he is drawing here is to the fact that you could not run a railway line with everyone running their own personal train (Latour 1996). You needed a centralized bureaucracy and effective monopoly in order to operate with any efficiency. Similarly, you could not have every train running on its own (local) time: you need a centralized time in order for the railway company to be able to create the representations that would permit efficient operation. Thirty-five years before the U.S. government recognized standard times across the United States, the railways imposed them in their own system (Cronon 1991, 79).

Chandler (1997), Campbell-Kelly (1994), and Yates (1989) have each chronicled the development of new office and accounting procedures in the railroad companies. All three chart how standardized annual reports, which represented the companies' operation within an absolute time and annihilated space (in Schivelbusch's sense) allowed the railroad companies to deal with the control and communication issues that arose. Thus Yates traces how the various genres of internal communication—the general order, the circular letter, the memorandum and the report—grew up in railroad companies (Yates 1989, 68). Her tracing of the development of the report form is indicative of the general movement. In the early days (the 1830s and 1840s): "railroad annual reports were generally designed as letters with opening salutations and complimentary closes" (ibid., 78). Thus they were historically specific events, tied to a given locale and addressed to a particular person. Over time, however, "tables would becomes increasingly important, and the letter form would disappear" (ibid.). The local dropped out and a regular *x* and *y* axis covering time and distance predominated. Reports became far more frequent too: the weekly report morphed into an hourly report marking the position of each train on the New York and Erie railroad. As Chandler (1977) notes, the new accounting forms developed by the railroads were adopted with minimal changes by the emergent new large-scale industries of the 1880s. Indeed, he asserts that they "remained the basic accounting techniques used by American business enterprise until well into the twentieth century" (117). Thus the new industries were made possible by the railroads and produced organizational representations modeled on the railroads. Small wonder that we can find features of the new dominant infrastructural technology so widely spread. Its organizational form (accounting techniques, reports, and so forth) synchronized with its own impact on the world (regularizing it)—providing both a material and metaphorical impulsion to order any particular form of enquiry or activity along these synchronized lines.

Technoscientific representations were socially and organizationally imposed by means of the new infrastructural technology—with a dual process of commodification and representation central to the shift. The same infrastructural technology that permits a qualitative leap in the process of commodification (the railway) also enforces a form of representation (abstract space and time) that is inherent in commodification. It enforces this form of representation not out of some kind of weird magic (or, worse, Hegelian dialectic) but for very good organizational reasons of control and communication. You need to be able to represent the world in a coherent and standard form in order to run railways and deal in commodities. Emerging here is Michel Serres's insight (1987) that since we live in a world where the human/nonhuman (nature/society) boundary is increasingly less well-defined, then we need analytic categories that allow us to account for the unified representational time and space applied to both bureaucratic and scientific work.

I have reproduced here some key findings from Alfred Sohn-Rethel's work on Galilean space-time. In his classic *Intellectual and Manual Labour: A Critique of Epistemology*, Sohn-Rethel (1978) focused on the relationship between the commodity form and the process of intellectual abstraction. His premise was that "the form of commodity is abstract and abstractness governs its whole orbit" (19). Thus, he continued, where use-value was concrete (one could use a commodity for a certain purpose in the real world), exchange-value was abstract, quantitative—reckoned in terms of the quintessentially abstract quality of money. Sohn-Rethel noted that when a commodity is up for sale, it is by definition not to be used; it exists in a kind of "frozen time" outside the normal flow of time. It moves in an abstract spatiotemporal world (which he calls—some score of years before Cronon—"second nature") unlike the concrete world of "first nature."

The commodity, then, sketches out (by moving in) a new kind of representative space. The links in his chain are the arguments that

> (1) Commodity exchange is "an original source of abstraction." The basic act of representation—separating properties of a thing from itself and charting those properties in a new medium (with its own time and space)—is a feature of capitalist organization. (2) This abstraction out there in the economic world (out there in second nature) "contains the formal elements essential for the cognitive faculty of conceptual thinking". Thus when people observe and describe the flow of commodities, they are in fact creating a representational space and time of much more general import. (3) This is more than a possible link—it actually describes the creation of "the ideal abstraction basic to Greek philosophy and to modern science." (Sohn-Rethel 1978, 28)

The concrete creation of a representational space and time comes first; abstract work by philosophers and scientists in this new space and time is consequent on that prior creation and epiphenomenal to it.

The act of synchronization brings material forms (commodity flow, train travel) together—and when, as we will now see, geological and historical time are mapped into the same time, powerful possibilities emerge for the metaphorical, ideological, and material to interact. And when they do, a new archive is created with the industrial age as the marker of a new time and space that henceforth will hold all records—be they business, scientific, or governmental.

Tales from the Past: Archives and Technology in the 1830s

In history, as in geology, a number of systematizers in the 1830s began to adumbrate a new kind of reasoning that operated in a flat space and time and which created a new kind of record of the past. For example, Michelet's systematization took the form of a nested series of regions that centered ultimately on Paris. The world was divided into several major climatic zones: "Follow from East to West the route of the sun and of the magnetic currents of the globe, the route of man's migrations, observing over this long voyage from Asia to Europe, from India to France, you will see that at each staging-post the fatalistic power of nature diminishes, and the influence of race and of climate become less tyrannical" (Michelet 1971, 229). This would be true for someone doing the trip today (synchronic extension) as well as for someone tracing the records of times past, where the Indian Empire preceded the

Egyptian preceded the Persian preceded the Western empires. A second world, Europe, is contained within this larger world: "Man has, step by step, broken with this natural Asian world, and constructed, through industry and trial a world informed by liberty" (ibid., 238). In so doing, He has created a second nature: "In general, after the great empires that lived according to nature, as one sees in Asia, have arisen states which are against nature, small, artificial states" (Michelet 1959, 229).

This European world has its own poles, its own India (Germany) and its own extension through time mirroring the macrocosm (Athens, Rome, Venice, Holland, the Hanseatic League). Nested within this world is the world of France, which operates the same regular spatial and temporal logic. Equally, within France is nested Paris—the greatest city, at the center of the world and riding the very crest of the wave of the present: "Germany has no center, Italy does not have one any more. France has a center; a single and unique center for several centuries—it has to be considered a person which lives and moves. The sign and guarantee of the living organism, the power of assimilation, is found here to the highest degree" (Michelet 1971, 247). A center of calculation indeed (Latour 1987)!

The great battle cry of the human spirit in this glorious race into the future was the destruction of nature: "With the world itself began a war which will finish with the world, and not before; that of mankind against nature, of spirit against matter, of liberty against fatality. History is nothing other than the tale of this interminable struggle" (Michelet 1971, 229). The grand determinant of his set of ordering devices for storing facts about the world (and without such a set of devices, histories are unruly, singular things) was the climate: "As you follow from east to west, along the route of the sun and of the magnetic currents of the globe, you see the migrations of humankind. And you see in it that at each station along the way on the long journey from Asia to Europe, from India to France, the fatal power of nature declines, and the influence of race and climate becomes less tyrannical" (ibid., 230). So we are being drawn into a land in which "temps" (weather) doesn't matter, and in which to a large extent "temps" (time) will not matter either. For in this world, history will become increasingly irrelevant. No room here for Herbert Read's suggestive "I freeze therefore I am an abstract painter" line of argument (1968). Nor will there be any environmental constraints whose tale must be told to understand the history. True liberty, for Michelet, does not have a past and does not have a grounding either on the face of the earth or in the present. It is fully emergent.

We will see this theme of the irrelevance of the past for a purified (real) present in Lyell's work later this chapter and in Comte's in the next. A similar theme can be found in Buchez's work. He argued that syllogistic reasoning from the Fall was pushing Christian people always back upon the past. He does not spell out the syllogism, but I suppose that runs something like this: "All individuals have original sin, society is composed of the set of all individuals, therefore society is subject to original sin." He depicted a world in which the timeless present would be free from the stain of past time:

> In society there is not, in reality, anything equivalent to what is called youth and senility in individuals: generations do not follow on one after another; all is mixed, in such a way that birth, death, adolescence, maturity, old age are always present at the same time, and in the same numerical proportions. It is a collective being destined to live indefinitely with an energy equal to what it had on its first day; for which the present is never anything, and for which the future is all; which is placed between a past that it continually leaves to advance towards a future that renews itself without end. Where to find an inexhaustible formula like social activity—a formula which never passes, and always contains within it an indefinite future. (Buchez 1833, 47–48)

His new formula (replacing the syllogism) was powered by the principle of the division of labor. This principle allowed acts that were carried out successively by individuals to be carried out in rapid succession—approaching simultaneity—of groups of people: "The succession of acts of which a single act is composed, is the same thing as the succession of diverse works necessary to arrive at a result which is nevertheless single. So this succession gives rise to what is called the division of labor" (ibid., 208–209). The division of labor used to lead to social inequality: "The order of generation in space becomes the order of subordination in time" (ibid., 209). However, in the (almost) eternal present of the well functioning society, things will be different: "The movement of this logic is so inherent to human nature, that it can only disappear with humanity; but one understands that it may in the future exercise itself almost simultaneously, in such a way that one ceases to see the other immense lacunae [the inequalities] we have been talking about" (ibid., 343). As society loses its mooring in the historical time of the Fall, the past (seniority) becomes effectively irrelevant.

For both these spatiotemporal systematizers and historians (Michelet and Buchez), as for Lyell and Comte as historians of science, the past as story would increasingly be a thing of the past. What should be recorded, increasingly, is not the context of any particular discovery or event. Thus for Michelet, when you get to Paris, upbringing and race are irrelevant to a person's actions, since all have the same upbringing and there is no difference between races—further away from Paris or further back in time, you need that information. Now, there will be no context—only text. Similarly, for Buchez, in earlier times, the principle of the division of labor constantly evoked the past in the form of seniority and power inequalities, whereas now the collective being of society will act effectively simultaneously and be constantly outside of time—it will never pass; it can only ever be present. The past could be generated from knowledge about causes such as climate or race; but contemporary humanity would move completely outside the flow of narrative time. The end of history, anyone?

The Memory of the Earth

In the 1830s, geology was a science in which, between the Plutonists belching volcanoes and the Neptunists spouting water, the catastrophists and the creationists had fashioned a history as a science of singularity and secular change (Porter 1977). The records of the earth were the records of catastrophic events that affected the body terrestrial; much as political events affected the body politic. Charles Lyell set out to refound this science—much as Buchez and Michelet refounded history—through a spatiotemporal systematization that reconfigured the world from a tapestry of tales to a random access archive.

Lyell saw himself setting down the basic laws that other geologists working empirically could draw on in their own studies. He gave a general rule that the kinds of forces acting in the world at the moment were the same kinds of causes that had always existed, at least as far back as the geological record went. This was a powerful rule. It meant that one could not refer back to a time when there were more earthquakes than at present, or when mountain ranges were thrust up in a single moment, and so on. One had to find slow-acting, steady causes in place of the "catastrophic" causes often referred to by his opponents, religious and otherwise. He offered one possible intellectual foundation for geology: the very title of his work echoes that of Newton's *Principia,* which was the paradigm case of a foundation text at the time Lyell was writing and which had done much to set the conditions for isotropic spatiotemporal analysis.

Geology was by far the dominant scientific discipline of the period. Aimé Boué, with the accounting passion so common at this time, summarized its phenomenal expansion in the following way: "Comparing the number of books published in 1833 to those in the years 1830, 1831 and 1832, the approximate proportion is established by the numbers 300, 450, 500 and 900" (*Bulletin de la Société Géologique de France,* 1833). In France, according to the *Écho du Monde Savant,* publications on geology and palaeontology in 1833 were far more numerous that those of all other sciences put together: "Physical and natural sciences (among them astronomy, physics, magnetism, meteorology, chemistry, hydrography and natural history): 144 books, 276 papers; palaeontology and geology: 61 books, 414 papers" (*Écho du Monde Savant,* June 20, 1834).

Everywhere one went in Paris, geologists were treating the glitterati and literati to their readings of the rocks of ages. The *Écho du Monde Savant* allows us to follow step by step the peregrinations of Parisian geologists:

In Paris this year Saturday and Sunday are essentially geological days. Saturday: at 9 in the morning, M. Brongniart begins his lecture on geological mineralogy at the Museum [of Natural History]; at 9 M. Boué delivers his private lecture in the rue Guénégaud; at 2 o'clock M. Élie de Beaumont steps on to the rostrum at the Collège de France; at 7 in the evening M. Boué delivers his public lecture at the Société de Civilisation; and at 8 o'clock M. Rozet starts at the Athénée. Sunday: MM. Constant-Prévost and Boué lead, separately, their troops, armed with hammers, canes and bags for rocks, rousing here and there the fear of the Republic or the edification of a school; while M. Boué explains, from 3 to 4, in the rooms of the Société, the geological relationships of the countries of Europe to those who, unwilling to expose their heads in the villages or their feet on bad roads, prefer to travel on the maps spread out for them by M. Boué. Nevertheless, it must be said that Saturday is going to lose M. Rozet and Sunday M. Boué. These two geologists finished their precious lectures last week, but in compensation M. Cordier's course, which will start soon, will offer a geological meeting and excursions with M. Élie de Beaumont. The organization of these will be announced soon, offering similar advantages on Sundays to a third band of rock-hunters. (*Écho du Monde Savant*, April 10, 1834)

There was a similar efflorescence of geological enthusiasm in England.

Geology was in the spirit of the time, and it had things to say about the time of the spirit. There are two types of time at work in Lyell. One is time as a passive container: it involves the attempt to give a chronology to the history of the earth, to trace its origin or to deny that there is any evidence that it has one. The second is time as process: it involves the attempt to pick out certain types of changes that are invariably associated with the history of the earth at any age and are thus in a sense a feature of time itself. Lyell addressed both religious time (sacred history) and human time (secular history) in such a way as to create a special time for geology that could be dealt with by professional geologists. Central to the understanding of the earth was an understanding of the nature of the archive. The heavily accented features of the earth were for Lyell a product of the way that the earth keeps its own records about itself and not a feature of variations over time in the constitution and virulence of its governing forces.

Lyell said that for all intents and purposes the earth could be taken as being eternal. It may once have had an origin, but no sign of this remains. This loss of the origin could be explained by the fact that the earth was molded by complementary destructive and creative forces. The latter (flowing water, tides and so on) visited each corner of the earth, grinding it down, dissolving it. The former (silt deposition, volcanoes, and so on) redistributed this formless matter, which thus bore no traces of its state before its dissolution. Each and every part of the earth only bears traces up to its last dissolution, and since there has been an indefinite number of these, there is no point in trying to discuss the origin of the earth. Lyell's geology has been taken as the triumph of "linear" time because it locates the earth along an indefinitely long line between the past and the future; however, beneath this crust of linearity we find a core of cyclical morphology for the earth.

In the following passages, we can get some picture of the workings of this calculus of temporal regularity:

There can be no doubt, that periods of disturbance and repose have followed each other in succession in every region of the globe; but it may be equally true, that the energy of the subterranean movements has been always uniform as regards the *whole earth*. The force of earthquakes may for a cycle of years have been invariably confined, as it is now, to large but determinate spaces, and may then have gradually shifted its position so that another region, which had for ages been at rest, became in its turn the grand theatre of action. (Lyell 1830–1833, 1:64).

In order to confine ourselves within the strict limit of analogy, we shall assume, lst, That the proportion of dry land to sea continues always the same. 2dly, That the volume of land rising above the level of the sea, is a constant quantity; and not only that its mean, but that its extreme height, are only liable to trifling variations. 3dly. That both the mean and extreme depth of the sea are equal at every epoch; and, 4thly, It will be consistent, with due caution, to assume, that the grouping together of the land in great continents is a necessary part of the economy of nature. (ibid., 1:112)

On this base, he argued for a climatic "great year"; the phrase is a reference to the Stoic's Great Year, which marked the period for the repetition of history. (ibid., 1:116)

We have now traced back the history of the European formations to that period when the seas and lakes were inhabited by a few only of the existing species of testacea, a period which we have designated *Eocene*, as indicating the *dawn* of the present state of the animate creation. But although a small number only of the living species of animals were then in being, there are ample grounds for inferring that all the great classes of the animal kingdom, such as they now exist, were then fully represented. (ibid., 2:225)

Species could, conceivably, survive complete "revolutions" of the earth's surface. (ibid., 2:225)

There is a consistent patterning to the disparate quotes of this text. In each, the part is taken as varying, as liable to be created or destroyed, whereas the whole is immutable and eternal—just like Buchez's society. Mediating between the two is cyclical change: a "cycle" of years attached to a region, a climatic great year attached to the earth over time, and "revolutions" of the earth's surface attached to species change.

The first volume of *Principles of Geology* gives a series of causes of change and shows how each destructive cause is equally, and in the same degree, constructive. Thus he writes with respect to sea currents: "In the Mediterranean, the same current which is rapidly destroying many parts of the African coast; between the Straits of Gibraltar and the Nile, preys also upon the Nilotic delta, and drifts the sediment of that great river to the eastward. To this source the rapid accretions of land on parts of the Syrian shores may be attributed" (Lyell 1830–1833, 1:308). Similarly, volcanoes on the surface of the earth seem to increase the general area of land mass, but submarine volcanoes raise the level of the sea, so the two cancel each other out (ibid., 1: chap. 18).

Lyell's assertion of a number of things that never change—land mass, degree of force of volcanic activity, and so on—seemed, in the opinion of his contemporaries, directly antithetical to the geological evidence. They also seem a long way from the kind of time we would expect to be associated with the Industrial Revolution, which was reaching its peak as Lyell wrote. The stillness, the ineluctable equilibrium between creation and destruction, contradicted the facts available to Lyell's contemporaries in various ways. One set of contradictions revolved around the whole schema, others around the position of humanity within it. In brief, the overarching problem was this: for all that Lyell might say that "present causes" explained all past geological occurrences, it was hard to believe that mighty mountain ranges were even now thrusting upward. Nature had left a series of monuments that looked for all the world like products of cataclysmic change of an order undreamt of today. Whole species disappeared in a flash from the fossil record. It scarcely seemed likely that massive continents had pushed out of the sea at an inch a century. Great truths demanded great causes. More probably, it seemed to most geologists, times had once been different, and the world was younger and more lively.

This image of an earth once lively going through a peaceful middle age was commonly used by Lyell's rivals, the catastrophists (Porter 1977). The catastrophic time that these geologists employed was used to reconcile the fossil and geological record with the Bible. The argument here was that it may seem difficult for all the evident changes on the face of the earth to have happened in six thousand years, yet what really happened is that time went faster then: there were more earthquakes, more volcanoes, and so on. This argument—a counterbalance to the generally perceived reality that human time was going faster now than it ever had in the past (a powerful myth still stalking our collective discourse today)—was used to give humanity a privileged position within the geological record. For, it was said, God waited until the earth was in repose before He introduced humanity—for whom it was created—onto its face. Lyell met both these privileged times head on in his work of defending the existence of a separate time for geology.

Lyell departed from previous traditions referring to the Book of Nature in explicitly developing the concept that the earth formed its own archive—though it was not a very good archivist. In so doing, he drew on the analogy of the information practices of statistics developed in large-scale government in the late eighteenth and early nineteenth centuries (cf. Hacking 1990; Foucault 1991). He noted that fossils were only created where new strata were being formed, and wrote the following:

> These areas, as we have proved, are always shifting their position, so that the fossilizing process, whereby the commemoration of the particular state of the organic world, at any given time, is effected, may be said to move about, visiting and revisiting different tracts in succession. In order more distinctly to elucidate our idea of the working of this machinery, let us compare it to a somewhat analogous case that might easily be said to occur in the history of human affairs. Let the mortality of the population of a large country represent the successive extinction of species, and the births of new individuals the introduction of new species. While these fluctuations are gradually taking place everywhere, suppose commissioners to be appointed to visit each province of the country in succession, taking an exact account of the number, names, and individual peculiarities of all the inhabitants, and leaving in each district a register containing a record of this information. If, after the completion of one census, another is immediately made after the same plan, and then another, there will, at last, be a series of statistical documents in each province. When these are arranged in chronological order, the contents of those which stand next to each other will differ according to the length of the intervals of time between the taking of each census. If, for example, all the registers are made in a single year, the proportion of deaths and births will be so small during the interval between the compiling of two successive documents, that the individuals described in each will be nearly identical; whereas, if there are sixty provinces, and the survey of each requires a year, there will be an almost entire discordance between the persons enumerated in two consecutive registers. (Lyell 1830–1833, 3:31)

Lyell observed that disease and migration might cause variance, and concluded: "The commissioners are supposed to visit the different province in rotation, whereas the commemorating process by which organic remains become fossilized, although they are always shifting from one area to another, are yet very irregular in their movements [so that] . . . the want of continuity in the series may become indefinitely great, and . . . the monuments which follow next in succession will by no means be equidistant from each other in point of time" (ibid., 3:31–32). This passage provides a litany of apparent disorder. It argues the uniformitarian case whereby, despite appearances of catastrophic change in the past history of the earth, the underlying reality was of incremental change. It was not that the earth was irregular, it was the earth's archival process that was less than efficient.

If the earth was a bad archivist, then it was up to the geologist to pull together information from a wide variety of sources so as to demonstrate the real regularity of its change. The earth's archival technology would have to be supplemented by an efficient use of what we now call information technology. Geologists in the 1830s saw this information-gathering effort as being heavily technologically mediated. Thus leading French geologist Léonce Elie de Beaumont gave a lecture in 1834 entitled "The Specialty of Geology deduced from the Special Nature of the Geologist's Life-Style." In his notes for the lecture, he wrote:

> The nature of geological science deduced from the order which establishes itself in the work of geologists . . . the geologist is therefore of all the classes of scientist the most obliged to displace himself . . . // that fact makes it even more likely to make him part of a distinct class than that this circumstance calls on a particular type of person . . . // of all the sciences, it is geology that relies most on improvement of the means of transport// means of transport are for the geologist what telescopes are for the astronomer. // the new roads that criss-cross Europe make the latter in some way a *geological preparation* . . . // remarks of Cuvier on steam boats//geology has in some way become a profession . . . where does geology begin and astronomy end? These two sciences are sisters and what above all places a line of demarcation between them is the different lifestyle they demand of their cultivators . . . one of the things which characterizes and even constitutes the progress of civilization is the division of occupations. . . . The establishment of railways will have the effect of enlarging geological localities, diminishing the distance between the geologists and the astronomer. (Beaumont 1832–1843, 1st lesson, December 7, 1839: f. 11)

Astronomy was in the nineteenth century the science of regularity (all the apparent perturbations in the earth's orbit were reckoned to be embedded in cycles of varying duration), so that the universe was effectively a clockwork mechanism. This was the dream of Laplace (1799) celebrated in the Bridgewater treatises, for example (Whewell 1833). It was also the ideal science, to which all others aspired. For Beaumont, then, the means to achieving the beauty and regularity of astronomy within the burgeoning science of geology was by going through the growing transport infrastructure. At the end of the day, when the roads had been well enough traveled, the two sciences would merge. He wrote that both specialties dealt with the following:

> periodic oscillations around a mean state. . . . However the heavenly bodies don't leave in space any trace of their passages . . . solar system a clock . . . the hands do not leave any traces. . . . As an archivist, the earth was a special kind of clock, an hourglass: This hourglass is the surface of our globe, and the scientists concerned with its functioning instead of calling themselves astronomers call themselves geologists. The objects of their sciences are contiguous and if the methods that they follow still separate them, one can say that they are sisters. The most remarkable thing about this hourglass is that

it preserves the traces of shocks it has received . . . can judge that their length is comparable to those of periods measure by the clock we were talking about. . . . Perhaps one day they could be linked and then the two sciences will help each other. (Beaumont 1832–1843, December 20, 1832: ff. 2–3)

Thus astronomy and geology will be seen to meet at the point of regularity, in a contemplation of the regular clockwork mechanism that seemed to govern both the heavens and the earth (all appearances to the contrary in the latter case; though a number of writers of the period also judged that human history was, in its underlying tendency, equally regular—e.g., Babbage 1837). At this point, first and second nature converge onto a timeless present and an anisotropic flow of time.

Technology entered into the picture of earth sciences in the 1830s in three ways then. First, the earth itself is a kind of large information storage device—not a very efficient one, but no less remarkable for all that. Second, the inefficiencies of the storage device could be mitigated—as Beaumont and Lyell both pointed out—by redundancies in the recording process; and these redundancies could be perceived through efficient use of the new transport infrastructure: steamboats and trains. Finally, when this work is done, geology would converge with astronomy in calling forth a historical time as regular and perfect as that of the clockwork solar system. If we had technology entering in at just one point of the process, then it could be simple analogy. What is significant is first of all the commitment to describing the archival process—a natural technology; then we get the move from the natural to the ideal archive being mediated by technology, with the ideal archive being marked by clockwork regularity. Beaumont and Lyell together embedded a complex argument on the nature of statistical record keeping, a central technology for a modern state, within their geologies. For both, good record keeping would demonstrate underlying lawlike regularities in the face of current empirical chaos: precisely the same argument being made in the then burgeoning field of government statistics (Porter 1986). The metaphor of the clock and its regularity supported the application of statistical thinking to the geological record; and the geological record thus sorted gave weight to the clock metaphor as organizing principle—each bootstrapped the other.

Through a linguistic metaphor, Lyell endeavored to explain the apparent asymmetry between past and present. This metaphor brings out the peculiar centrality of humanity in Lyell's geology and thus the centrality of human society to his problematic. It revolves around an image sanctioned by long usage in scientific texts: the idea of the Book of Nature. In the natural theology that Lyell opposed, the Book of Nature was taken to be fully complementary to the Book (the Bible). In his alternative development of the theme, the miracle would be if we *could* read the Book now, since we only know a tenth of the world in the present and so could only know a tenth of it in the past. He wrote playfully:

> So if a student of Nature, who, when he first examines the monuments of former change upon our globe, is acquainted only with one-tenth part of the processes now going on upon or far below the surface, or in the depths of the sea, should still find that he comprehends at once the imports of the signs of all, or even half the changes that went on in the same regions some hundred or thousand centuries ago, he might declare without hesitation that the ancient laws of nature had been subverted. (Lyell 1830–1833, 1:462)

The logic of this passage is not, perhaps, immediately clear; not surprisingly, it was dropped from later editions. What Lyell is saying is that at present our knowledge of the Book of Nature is highly restricted (to processes occurring on land, and only to a small proportion of these). He argues that if from our knowledge of these processes we could reconstruct the history of the earth, then the past must have been very different—for that would mean that the small proportion of causes that we know about today were once all the causes there were. Whereas the earth kept only a limited and random sample of its own records; here we as humans have access only to a limited and random grammar of the Book of Nature. So the burden of proof lies not with those who with their set of present causes cannot explain the past, but with those who with their set of past causes can.

Lyell's defense of his geological time against appearances to the contrary is, first, that appearances are necessarily deceptive if his system is right; and second, that there is no way at present that geologists could know enough to explain past changes. He argued against any possible connection between religious time and geological time by denying an origin to the earth and buttressed his new geological time against possible counterarguments about the nature of the geological record. When he arrived at this point in his argument, he believed that the basis had been laid for a true science of geology—an argument he proposed using the contrast between his own true language of geology and the false language of catastrophists: "These topics we regard as constituting the alphabet and grammar of geology; not that we expect from such studies to obtain a key to the interpretation of all geological phenomena, but because they must form the groundwork from which we must rise to the contemplation of more general questions relating to the complicated results to which, in an indefinite lapse of ages; the existing causes of change may give rise" (Lyell 1830–1833, 3:10).

He made two further moves in order to defend his time. First, he tried to legislate for the way that geology would develop as a discipline by trying to attach to it the same time that he attached to the history of the earth. Second, he produced arguments to counter the idea that geological time was somehow different since the advent of humanity—more peaceful, or transformed by its presence. Just as we would not contribute henceforth as geniuses to the scientific record but as workers in the mines; so humanity did not contribute anything lasting to the earth's archives by virtue of our consciousness. Overall, the appearance of any discontinuity in the past—of a science, of the earth—was to remain a feature of the imperfect record, not of reality.

For Lyell, just as the past history of geology is concerned with catastrophes, so is the past history of the discipline of geology catastrophic; it is "between new opinions and ancient doctrines, sanctioned by the implicit faith of many generations, and supposed to rest on scriptural authority" (Lyell 1830–1833, 1:72). Lyell does not, however, abandon his patterning for geological time when he turns to geologists. The imperceptibly slow operation of simple causes operates for both the earth and its scientists: "By the consideration of these topics, the mind was slowly and insensibly withdrawn from imaginary pictures of catastrophes and chaotic confusion, such as haunted the imagination of the early cosmogonists" (ibid.). To get some idea of just how long a period of time he has in mind, we can turn to his proto-Jungian assertion that "the superstitions of a savage tribe are transmitted through all the progressive stages of society, till they exert a powerful influence on the mind of the philosopher" (ibid.). Thus the catastrophic history of geology is itself underwritten by slow, insensible change. The two rhythms of time (the catastrophic and the uniformitarian) battle it out within both the history of geological ideas and the history of the earth. Just as our readings of the Book of Nature should become ever more uniformitarian, so should our reading of the history of geology. Lyell signals this change in the nature of the history of geology in a return to the language metaphor. In a lecture to London high society, he referred to the former (catastrophic) state of geology:

> While the science was in so fluctuating a state the philosopher who was anxious to discover truth, naturally preferred to enter himself into the field of original investigation, rather than to devote his literary labors; to the comparison and the reduction into order of imperfect observations and a limited collection of facts. One of our poets alluding to the incessant fluctuations of our language after the time of Chaucer complains that:
> "We write on sand, the language grows
> And like the tide our work o'erflows." (Lyell 1833)

This was in marked contrast to the halcyon future, when, as Lyell (ibid.) opines: "We shall from year to year approach nearer to the time when the new facts which can be added by one generation of men however important will form but a trifling contribution to the stock of knowledge which had previously been acquired and when that period shall arrive they who have no opportunity of traveling themselves or of constantly associating with those who are engaged in actual observation will be more on a par." When the center of calculation is in place, there is no need to move. So static scientists will be able to produce science about an unchanging earth: their records will last forever through archival publications because they will be reading the true records of the earth. Sublunary earth would become a part of the stasis and regular order that exists everywhere but, apparently, here.

We can, then, unify Lyell's pictures of the history of geology and the history of the earth. In the past, knowledge developed catastrophically, and analyses were framed in terms of catastrophes; in the present and future, knowledge develops uniformly and analyses are framed in terms of continual steady change. Lyell encourages us in this formulation when he asserts that "the connexion between the doctrine of successive catastrophes and repeated deteriorations in the moral character of the human race is more intimate and natural than might at first be imagined" (Lyell 1830–1833, 1:10). This is the same move that Buchez made: science would be the agent that converged both human and natural history onto isotropic time. There is, indeed, a powerful moral force in Lyell's geology that derives from just this symmetry between the past of the geological discipline and of the earth. It would be better all round, the reader feels, if the time that has eternally framed nature were to frame human society. It is through records that salvation will be found—scientific records (properly kept) tie us in with the (real) records inscribed onto the skin of the earth (Derrida has a nice discussion of the inscription of records onto skins—in his case, Freud's penis (Derrida 1995)).

There is a final way in which the new time that Lyell is using to found the discipline of geology is applied in his *Principles of Geology*. This is in his resolution of the problem of whether the time of the earth is somehow different since the creation of humanity. Unlike all the other objects in Lyell's geology, humanity irrupts into the picture at a very specific moment. Moreover, this moment is six thousand years ago: precisely the moment that biblical fundamentalists picked for the origin of the whole earth (including humanity). Not only did humanity make a singular appearance, however, it also set about creating the appearance of singularity. Thus Lyell (1830–1833) commented on hybrids that displayed extreme variability in their outward form (and thus changed at a pace too fast for his geology) and asserted that no hybrid had ever achieved a permanent niche on earth (ibid., 2:32 and 3: chap. 4). Humanity, then, makes time look as if it is irreversible and rapid (even catastrophic), but for Lyell this serves only to highlight the fact that underlying reality is as uniform as can be. In general, not only does humanity have a tendency to read the Book of Nature wrongly, it also has had a tendency to write it wrongly too, making the same mistake in each instance.

Lyell has two strategies for playing down humanity's influence: accreting it to the natural, and assigning it to another plane of existence. In the former, Lyell stresses that changes wrought by humanity are for all that natural changes. Humanity does its work of sowing seeds far afield, but these seeds would have been sown regardless: by the wind or through the agency of a migrating bird. Nature keeps a check on the whole process by organizing flora and fauna into "nations": nothing can survive long outside its nation. This "natural" side of humanity is totally divorced from its civilized side. Thus Lyell (1830–1833, 3: chap. 5) argues that if humanity were cataclysmically reduced to a few specimens in some far-off land, it would once again spread out and fill the earth because that is our natural fate—whether or not we are civilized. Both the spread of humanity and its ability to act as dispersive agent are, then, fully natural and under nature's control.

There is, however, another aspect to humanity: its ability to transform landscapes and species temporarily. To account for this aspect, Lyell develops his second strategy for playing down humanity's influence: he posits a complete divorce between civilized humanity and nature. The changes humanity has wrought are

> not of a *physical* but of a *moral* nature . . . it will scarcely be disputed that we have no right to anticipate any modification in the results of existing causes in times to come, which are not conformable to analogy, unless they be produced by the progressive development of human power, or perhaps from some other new relations between the moral and material worlds. In the same manner we must concede, that when we speculate on the vicissitudes of the animate and inanimate creation in former ages, we have no ground for expecting any anomalous results, unless where man has interfered, or unless clear indications appear of some other moral form of temporary derangement. (Lyell 1830–1833, 1:164)

The two arguments about human time can be summarized thus: insofar as humanity interacts with geological time, it is the bestial part of humanity fitting into the economy of nature (a phrase much used by Lyell), whereas civilized humanity operates in a different dimension to nature and creates the temporary appearance of an anomaly in Nature's Book. We humans do not contribute to the "real record." So the earth creates a false record by being a bad archivist; and the humans create false records through smudging the earth's annals with mules, hybrids, and monsters. But the geologist can read between the lines to the true record of stasis. There is a similar consideration of humanity's effects in some contemporary environmental literature—where humanity's influence is seen as ultimately outside of the natural system. In both cases, this allows the two "morals" of Lyell (state of morality and scientific theory earlier; human action and state of the earth here) to converge and hence intercommunicate along a Möbius strip).

In general, Lyell's foundation work on the creation of geological time operates a series of divorces. The time of the origin is given to religion, the rest of time (effectively all of time) is given to the geologist. Catastrophic change is given to the history of the earth sciences before the foundation of geology by Lyell; the new discipline of geology will be uniformitarian. Humanity's "moral" influence is seen as outside geological time and fully reversible; its "physical" influence is fully within geological time. Thus a single time is created for the history of the earth, for the development of earth sciences and for human time—and it is the time of the good record keeper.

Creating a Memory for Geology

A series of tracts on natural theology were published in England during the 1830s, tracts called the Bridgewater Treatises. These formed a major series of books written by leading scientists of all disciplines. They had been commissioned in a bequest made by the dissolute ninth Earl of Bridgewater. The earl made his fortune in building canals in the industrial north of England, but he squandered most of his money with sybarites and sycophants. In his will he made provision for the publication of a series of pious works, as a gauge against his entry into Paradise. The President of the Royal Society, with the help of the Bishop of London and the Archbishop of Canterbury, chose eight leading men of science, who were instructed to shed light on "the power, wisdom and goodness of God, manifested in his Creation, illustrating the proof with all reasonable arguments." Charles Babbage, who invented an ancestor of the computer, wrote a ninth "renegade," uncommissioned treatise. The sort of reasoning proposed over the set of treatises was that the existence of God was proven by the fact that all of Creation formed a perfectly ordered whole, with even—indeed especially—the anomalies appearing as the expression of intelligent design. Thus water is paradoxically heavier as a liquid at 4 degrees Centigrade than as a solid, so that ice floats on water and fish can survive in lakes through winter (Prout 1834, 251). Though they were supposed to consider all the sciences, most of them contain long chapters devoted to geology. The Bridgewater Treatise by the Reverend William Buckland was devoted to geology; it provides a model of the religious causality that Lyell was fighting. Babbage's work is devoted to the subject of physical causality. For both Buckland and Babbage, the argument that Lyell made—that the archive is jussive and sequential—is central.

Buckland's geology represented the opinion of High Church religious authorities (as against those of the fundamentalists; many in the Low Church, even later in the century, cleaved to a literal reading of the Bible). We will see that clergymen were prepared to follow Lyell to a certain point but without going far enough to raise the question of the infallibility of the Bible or the Book of Nature. Buckland made frequent allusions to the work of Lyell and admitted the principle of the earth as having persisted through an indefinite number of ages. He had two methods of squaring Lyell with the Book of Genesis. The first, which has survived to this day, is to say that while a rose is but a rose, a day can be any number of things—extending to hundreds of thousands of years if God saw fit. (Le comte Charles de Perron ([1835] 1840, 13), pointed out that we sought to impose our ideas of time on God because He existed outside of time and so had no such conception.) The second was that he found a loophole in time in Genesis in the undefined interval following the first verse—so that in this period fossil evidence and so forth could accumulate, and it was only with the third verse that the current sequence started—the requisite six thousand years before. Genesis, then, is literally right, and so is Lyell . . . about the age of the earth. The records in the Book of Nature thus reconciled with the Book itself (Buckland 1836). Whereas for Lyell nature is profoundly and perhaps irretrievably unknowable, for Buckland it is all in essence already known. Buckland's infallible Book of Nature contains the sure traces of God's design, which provides the true link for geological events; Lyell's fallible book obscures these very traces.

Indeed, Lyell was angry at his friend Babbage's use of Laplacean determinism, with its entirely knowable past and futures precipated onto the moment of calculation of the scientist in the present (or God whenever, depending on your preference). Laplace (1814) had written the following: "An intelligence who at some given moment knew all the forces that animate nature, and the respective situation of the beings that compose it, if it were further sufficiently vast to submit these data to analysis, could embrace within a single formula, the movements of the largest bodies of the universe and those of the lightest atom: nothing would be uncertain for it, and the future, like the past, would be present to its eyes" (2–3). Babbage's ninth Bridgewater Treatise made much of this kind of determinism, which mimicked the operation of his calculating engines, which could even be programmed to contain the numerical equivalent of miracles, if the algorithm were complex enough. The Difference Engine could perform the same action without change for 100,000 years, and then produce an anomalous result (a "miracle") and immediately return to normal. The fact of the anomaly was not enough to deny the regularity of the machine. This determinism not only applied to the future, but could also in theory be used to learn about the past. He wrote that "the air itself is one vast library" (Babbage 1837, 113) because when we speak "the waves of air thus raised, perambulate the earth and the ocean's surface, and in less than twenty hours every atom of its atmosphere takes up the altered movement due to that infinitesimal portion of the primitive motion which has been conveyed to it through countless channels, and which must continue to influence its path throughout its future existence" (ibid., 110). He believed that if we knew the original position of every atom in the atmosphere, we could trace its complete future. Every murderer bore a record of his crime, "some movement derived from that very muscular effort, by which the crime itself was perpetrated" (ibid., 117). One great dimension of the past was the destructive nature of a totalizing social or natural memory (remembering murders, brutality, and missteps)—and indeed his proffered punishment for some crimes was hearing the past repeated (see Liu, forthcoming, for a very suggestive analysis of a negative memory passage) (ibid., 164). Thus he shows how even minute phase differences of tide in a spheroidal world with two great tides would have a large effect over hundreds of years (ibid., 248). He is in horror of these irregularities. The incalculable, irrational past seems to him full of noise and cacophony—irregularity, just as for Lyell the past of geology was cacophonous. A good enough ear could hear the Sermon on the Mount, without making Monty Python's error of believing that the Greeks would inherit the earth. But the use of such an instrument would basically be torture; a future punishment would be the connecting of the soul of a dead man to a "very sensitive bodily organ of hearing." Suddenly "all the accumulated words pronounced from the creation of mankind will fall at once upon that ear." This repeated image of past noise is complemented by a drear statement that for most "oblivion would be the greatest boon" (ibid., 165) and a look forward into a well-ordered future: "if that Being who assigned to us those faculties, should turn their application from survey of the past, the inquiry into the present and to the search into the future, the most enduring happiness will arise from the most inexhaustible source" (ibid., 166).

This contrasted with the memory ensconced in books (filtered memory; rational memory; part of the archive with its very clear point of origin in printing) that freed us from instinct and brutality. Thus Babbage (1837) wrote that, until the invention of printing, "the mass of mankind were in many respects almost the creatures of instinct" (59). Now, the great are encouraged to write, knowing that "they may accelerate the approaching dawn of that day which shall pour a flood of light over the darkened intellects of their thankless countrymen" (ibid., 55), seeking "that higher homage, alike independent of space and time, which their memory shall for ever receive from the good and the gifted of all countries and all ages" (ibid., 54). Since printing, the rate of progress of humanity has "vastly accelerated" (ibid., 55); over the past three or four centuries, "man, considered as a species, has commenced the development of his intellectual faculties" (ibid., 56). In order for this the new space and time, he needs the regular working of the machines of nature and the world. The act of making information such a key variable tied directly into operations on social and natural space and time: for Babbage, the regularization of time and the distribution of tasks. In both cases, the mythological operation succeeded because of infrastructural work: for Babbage the development of computing and other machinery and the imposition of the principle of the division of labor.

Lyell, on receiving a first manuscript of this book from Babbage, fired off a series of criticisms of these passages. Basically these consisted, as in his criticism of design, of stressing over and over the fallibility of the Book of Nature. Thus he wrote to his friend:

> if it be true that all sounds remain in the air, which I cannot help doubting, something should be said for the benefit of the ignorant . . . Can the air be said to be the historian when it is only a mute depositary unread by any one and unheard? Do not the circles on the water cease at last, an ordinary reader (for whom you write) will feel annoyed at not being told how it is that in a resisting medium undulations are not at length destroyed, how it is that they do not combine with others so as to produce new sounds and notes and words? (Babbage 1830–1840, May 1837, f.187)

In general, he considered the book in bad taste, and recommended against publication.

Babbage's is certainly an extreme expression of the theme of complete determinism, but the trope was common at the time. Buckland and Babbage proffered two arguments for the complete knowability of the Book of Nature: one threatening to subsume geology into theology, and the second threatened to subsume it into physics. The Word, lodged in the Ark of the Covenant; or the World, lodged in the Archives of the Exact Sciences, were equally corrosive of Lyell's archive; the geologist needs to read the smudged archives of the earth.

Lyell created a picture of the work of geologists that allowed him to know nature without being a theologian or a physicist. He created a "Mnemonick Deep" that anchors us in an eternal now. For Lyell, the role of the interpreter of nature is central: God and Nature are both profoundly unknowable, and it is only through an epiphanic moment of profound insight that the scientist can hope to grasp their mysteries. Lyell (1830–1833) referred to this moment in a citation from Niebuhr: "'he who calls what was vanished back again into being, enjoys a bliss like that of creation'" (1:73). Any human attempt, religiously motivated or not, to offer some more direct way of reading or writing the Book of Nature deserved utter scorn. Along with the Romantic poets (e.g., Keats in his "Ode on a Grecian Urn" or Byron in "Childe Harolde"), Lyell found reverence and sublimity in the capture of the tension between the fleeting instant and eternal ages; thus anyone who saw "the summit of Etna often breaking through the clouds for a moment with its dazzling snows, and being then as suddenly withdrawn" must "form the most exalted conception of the antiquity of the mountain" (ibid., 3:370). For some, size matters; for others, it is age and beauty.

The Profession of Geologist

The agents of the convergence of human and earth time onto a stateless present were readers of the archives of the earth, the geologists—workers in the leading science of the time. Not so surprising, perhaps. If one looks at the role of information and database theory in genomics, a core science of our day, one sees a great deal of continuity over the past few hundred years. The archive is central both to thinking about the objects being studied (the annals of the earth; the code in our genes) and to our writing these studies (archival publications; databases as scientific publications).

Links between the information explosion of the early nineteenth century and the new ways of describing the past of the earth were forged (if that is a good word) explicitly in the jottings of Léonce Elie de Beaumont. His notes for an introductory lecture on geology at the Collège de France in 1839 include the following:

> today, now that we start to be able to go to St. Petersburg in 5 days, to Constantinople in 8 to 10 and to New York in 14, given that today with the electric telegraph people talk to each other by signs at several hundred leagues distance; we are at the start of a new era when the locality of each person will be much bigger than it has been up to now because the ability to move around will have been much increased and the inconvenience of being away from home will be greatly diminished. We are approaching a time when the locality of each geologist will be the terrestrial globe. It is then that a philosopher will be really able to call himself citizen of the universe. (Beaumont 1832–1843, 1st lesson, December 7, 1834: f.11)

> Buffon ended the heroic age of geology wherein everyone constructed complete systems—it was impossible to go further without making geology the province of a large number of people and as a consequence a profession having its own rules . . . it is after him and not though him that geology took its place among the academic sciences, which grow gradually through the successive works of a collection of individuals, it is the application of the principle of the division of labour. (ibid., ff:14, 15)

Lyell himself made much of the need for travel as being central to the occupation of the geologist; indeed, in his autobiography he proclaimed that: "We must preach up traveling as the first, second, and third requisites for a modern geologist" (Lyell 1881, 273).

Beaumont promoted the principle of the division of labor in modern geology. This ties in in several ways to the idea of specific geological time found in Lyell's *Principles of Geology.* Concentrating for a minute on what is common to both authors, we find that they both make exactly the same points about the development of geology. Now is the end of the "heroic age," of individual systems that are thrown up and hurled down in cataclysmic succession. For both, the present is the time of slow, piecemeal development by a large group of workers, no one of whom will dominate the field, and both authors in their work laid quasi-mathematical foundations for this field: Beaumont with his theory of orogeny, and Lyell with his balance sheet for the static earth. Lyell spent large sections of his *Principles of Geology* inveighing against what could be called the "heroic" system of geological change. There was no time when things were different: "The minute investigations . . . of the relics of the animate creation of former ages, had a powerful effect in dispelling the illusion which had long prevailed concerning the absence of analogy between the ancient and modem state of our planet" (Lyell 1830–1833, 1:72).

Why this division of labor, within geology as a discipline, and in society and nature? Both Lyell and Beaumont stress that, with current social and economic change, they were seeing a form of information explosion. If we look at Lyell's geology as a system for the classification of this information, then we can gain another insight into the articulation of his time. At its most abstract, Lyell is proposing a change from seeing geology as a litany of an enormous number of singular events (like a huge epic poem) to seeing it as the systematization of a small number of kinds of event. Thus, instead of seeing a particular mountain as a sign of a massive upthrust at some given date in the past, he sees it as a typical example of a kind of change that is occurring today. There are no privileged moments. His geology is a kind of bookkeeping device that allows the storage of vast amounts of information by sorting them into a kind of filing cabinet of different kinds of event. This reading of Lyell brings out why it was easy and natural for Lyell to find the metaphor of the statistical commissioners—after all, his geology is undertaking a version of their task. Further, it demonstrates how large-scale social change is reflected directly in the writing of geology through the intermediary of the organization of the foundling discipline of geology (the principle of the division of labor) and the handling of the information explosion that all sciences and professions were undergoing (uniformitarian time).

Indeed, Lyell's *Principles of Geology* reads like nothing other than a double-entry ledger-book: the sum of creative and destructive forces (credit and debit) is always precisely zero. Lyell carries this principle well beyond the bounds of the available evidence in his four rules of the disposition of land and sea, which we cited previously in this chapter. To recapitulate, these were that the proportion of dry land to sea is always constant, that the volume of land rising above the sea is constant, that the mean and extreme depth of the sea are equal at every epoch, and that "the grouping together of the land in great continents is a necessary part of the economy of nature." These rules are frankly absurd unless they are read in the context of Lyell's accounting method. A further justification for the reading lies in Lyell's constant reference to the economy of nature, the plan of nature. Thus it helps us interpret the following enigmatic opinion about the idea some philosophers had that only a few laws produced the "endless diversity of effects": "Whether we coincide or not in this doctrine, we must admit that the gradual progress of opinion concerning the succession of phenomena in remote eras, resembles in a singular manner that which accompanies the growing intelligence of every people in regard to the economy of nature in modern times" (Lyell 1830–1833, 1:76). The metaphor of the economy of nature is second in his work only to the Book of Nature. Lyell introduces a principle of the division of labor into the profession of geology and into the economy of nature, and such that both will generate cumulative reports in a stateless present. The earth is an open book if the principle of the balance of forces is accepted: it returns a calculus of regularity out of apparent chaos and old time.

Lyell in his *Principles of Geology* (1830–1833), as we have seen in Beaumont's work, drew a close connection between the work of the geologists and the work of the astronomer:

> However convinced a geologist may be that the earth had a beginning he has no right to assume a priori that in tracing back the history of the globe he should find the records of that beginning, no more than we have a right to assume in regard to any particular nation that we shall be able to trace back their history to its true origin. . . . When Descartes removed the boundaries of universe and speculate on indefinite space as filled with worlds, no one had a right to impute to him that there was no termination to the space. (August 7, 1811, f.6)

Just as the astronomers had reduced the solar system to a clock, then so could geology reduce the earth to an hourglass. This is precisely what Lyell (1830–1833) did in his articulation of time: every physical operation is made as regular and smooth as clockwork; it was just a case of finding the right periods. Thus he recognized one difficulty with his system: "It is clear that if the agency of inorganic causes be uniform as we have supposed, they must operate very irregularly on the state of organic beings, so that the rate according to which these will change in particular regions will not be equal in equal periods of time; nor do we doubt that if very considerable periods of equal duration could be taken into our consideration and compared one with another, the rate of change in the living as well as in the inorganic world, would be nearly uniform" (ibid., 2:160). The solar system was hymned as an accurate clock, the clock dominated industry, and between the two the synthesizing geologist Lyell turned the earth itself into a clock: ticking away regularly and faithfully once we understand its workings. Lyell's articulation of the connection between geological and human time can be interpreted in this light. Humanity and geology may have developed raggedly in the past, but with the triumph of industrial society (which was the natural form of association because it ran like clockwork) the two could approximate to the industrial time written into his geology. Thus two basic methods of factory production—the division of labor and the parceling up of time into regular units—are both written into the time that Lyell created for the new discipline of geology; and we have seen that he used factory production methods precisely because he saw these as best suited to the fruitful exercise of the profession he sought to create. Through this approach, social time could converge with natural time; astronomical time with geological time.

This is a long way from the picture of Lyell as the heroic scientist who rolled back the years of the earth's origin. Keeping to this one canonic result, there would be no way to situate his work within the Industrial Revolution or to see its link with the Romantic movement sweeping Europe at this time. In particular, the thrust of what he says about the nature of time—his stress on the periods of equal duration that govern industry, the solar system, and the world; and his belief in an epiphanic moment wherein all of time is grasped—would have been missed.

The jussive work of Lyell's archive, which had him expelling the priests from the temple of science by redefining geological time has become clear. In fact, two sorts of religious time are excluded from geology by Lyell. The first is the pagan representation of a time of great heroes bestriding the field of geology like colossi; or of great geological events—earthquakes, floods, storms—dwarfing today's minimal, tranquil variations on the theme of repose. The second is the Christian author of the Book of Nature being denied the right to interpret His works: it is the moment of creation enjoyed by the geologist as the new priest of nature that constitutes the definitive, correct reading of the flawed Book. Whereas God had been the only being capable of standing outside time and space and able to oversee the whole, now the geologist could join and effectively supplant Him. The foundation myth is thus shorthand for a much more complex reality that sees modern science, despite its secular self-image, forming itself into the new religion of our times.

So does our hero, Charles Lyell, wield enormous power? Armed only with his incisive intellect, he singlehandedly engineered the split of Church and State so as to found the profession of geology? Of course, this vision is totally improbable. The arrow of historical causation in this case is not from towering intellect to society through the mediation of ideas, but from society to intellect mediated by the day-to-day exercise of the profession of geology. The problem of the division of labor and the organization of time in factories and in geology was precisely the same problem. Through the mediation of the creation of the profession of geologist in the image of the middle management of a thriving business, Lyell inscribed the same time scientifically onto the history of the earth as others inscribed socially onto industrial society. He produced in his *Principles of Geology* a reading of geological history and of the earth that revolved around an understanding of archives. It is scarcely surprising that Lyell uses the metaphors he does and Beaumont makes the connections he does: both were being better historians than an intellectual historian who asserts that all Lyell did was increase the age of the earth.

Conclusion

Synchrony and Synchronization

We have had three "second natures" in this chapter. One was from the past, and two are analytic constructs from the extended present. Ajasson and Grandsagne had industry as second nature; Sohn-Rethel had smooth, isotropic space and time in that position; and Cronon had the imposition of a Cartesian grid on the grand prairies of the Midwest. We also encountered Michelet's "second world," "constructed . . . through industry and trial." All are speaking of the moment of the penetration of the industrial economy into the modern world over the span of several centuries.

One way of understanding this "second nature" is that before it, things were lumpy, confused, wayward. True science, Lyellian geology, stood on the leeward side of a chaos characterized by vast systems erupting into the world of discourse and then disappearing abruptly; vast causes acting punctually in the history of the earth and then disappearing. Humanity for Babbage had created a new human nature on the leeward side of a chaos characterized by an irregular past and a totalizing memory. On this side lay the orderly production of the scientific archive—remembering for us only what needed to be remembered and consigning chaos to oblivion. One could guarantee against chaos and old time on two fronts: by keeping better records (Babbage, Lyell) and by better reading the earth's records beneath their chaotic inscription on the face of the earth. Our moral duty as humans was to recognize the stasis at the heart of disorder (apperceive synchrony in the midst of diachrony) and to bring social and natural time into a unified form (synchronization) through the production and storage of records.

The trick that Michelet and Lyell and de Beaumont deployed to make the earth's, humanity's, and geology's records more regular was to spatialize time. For the earth, every cause that has ever acted could be found acting somewhere in the world now; every story that appeared wayward and secular could become part of a metronomic story if due attention were paid to the earth's record keeping and if the true story were garnered from the set of imperfect traces scattered across the face of the earth. For humanity, going out in space (to India, to the Orient, or within France to Brittany or the Pyrennees) was going backward in time. The mapping process allowed the regularization of the earth's history. For geology, the past was catastrophic, but progress would become regular as the set of geologists across the face of the globe replaced the individual genius irrupting on the scene toting a completely new theory.

In the process of spatializing time, humanity would be increasingly written out of the story of our planet. For the earth, this meant that our impact on the surface was anomolous, moral, and short term. For humanity, the surface's impact on us was progressively less the closer we got to Paris—so that in the City of Light, weather and topography were irrelevent. For geology, the individual, named scientist would be replaced by the faceless scientist working in a vast and effective machine. Engineering this removal of humanity from the face of the earth was the Industrial Revolution. For our understanding of the earth, it was our ability to travel by steamboat and train that allowed us to bring the sister sciences of astronomy and geology into synch. For the history of humanity, it was the ability of industrial artifacts to shelter us from external influence. This paradoxical removal of humanity from the face of the earth just as we were achieving an apogee of impact has had a long history through to the present—the concept of "ecosystem," so influential in ecology, was largely undergirded by the idea that we humans were not part of natural systems (O'Neill 2001; cf. chapter 5).

Memory Modalities

New memory practices, of the sort we have explored in this chapter, lie at the heart of our ways of knowing both ourselves and the world. They skew our available ontological space. From one point of view, our contemporary memory regime seems to be a highly prosaic affair involved with the growth of computers and their associated bureaucratic structures. From another perspective, and often from the same writer within the same paragraph, it seems to be a revolution dense with meaning that will unlock the secrets of life and the universe. Beniger's *The Control Revolution* is a prime example of this double trend. Thus he writes very prosaically: "The rapid development of rationalization and bureaucracy in the middle and late nineteenth century led to a succession of dramatic new information-processing and communication technologies. These innovations served to contain the control crisis of industrial society in what can be treated as three distinct areas of economic activity: production, distribution, and consumption of goods and services" (Beninger 1986, 16). At the same time, on the same page, he can pose a question like "Why has information—of all commodities—come to dominate the economies of at least a half-dozen advanced industrial nations?" (ibid., 58). He answers this and like questions with the following: "Ultimate answers to these questions, we have found, lie at the heart of physical existence. In order to oppose entropy and put off for a time the inevitable heat death, every living system must maintain its organization by processing matter and energy. Information processing and programmed decision are the means by which such material processing is controlled in living systems, from macromolecules of DNA to the global economy" (ibid., 58–59). The nature of existence and the new technology are resonant one with the other.

Or again, Herbert Simon tells us in the space of a paragraph: "From an economic standpoint, the modern computer is simply the most recent of a long line of new technologies that increase productivity and cause a gradual shift from manufacturing to service employment. . . . Perhaps the greatest significance of the computer lies in its impact on Man's view of himself. No longer accepting the geocentric view of the universe, he now begins to learn that mind, too, is a phenomenon of nature, explainable in terms of simple mechanisms" (qtd. in Forester 1980, 434). This last point, echoing Wiener's sketch of the history of science (humanity being no longer spatially then temporally then intellectually central), places the information revolution on an eschatological base at the same time as the first point puts it on an economic base (Wiener 1951). *The Oxford Dictionary of Computing* (Illingworth 1983) ranges widely in two consecutive sentences: "In principle any conceivable material structure or energy flow could be used to carry information. The scale of our use of information is one of the most important distinctions between the human species and all others, and the importance of information as an economic commodity is one of the most important characteristics of the 'post-industrial' civilization, which we are often said now to be entering" (see under "Information"). From the nature of the universe to the nature of humanity to the organization of the economy in three easy steps. These three steps were mediated in the nineteenth century, in the dominant science of the time, by a meditation on record keeping.

Perhaps the most surprising thing about these conjunctions is that they do not surprise. We are used to the mythological dimension of new memory practices. Business history gets regularly articulated with the meaning of life in a nonproblematic way. One possible treatment of this articulation is to dismiss it as hyperbole. For its detractors, this is the simplest and perhaps the most common treatment, but it is also one that denies the complexity of this historical phenomenon. There is a compelling connection between the information revolution as an economic fact and as a statement about the nature of the universe. This information mythology stems from a set of work practices whose constitution illustrates an important dimension of the relationship between information, knowledge and society—one that drives to the heart of our political economy: the interface between the social and natural worlds.

In information mythology, "information" can travel anywhere and be made up of anything. Sequences in a gene, energy levels in an atom, zeroes and ones in a machine, and signals from a satellite are all "information," subject to the same laws. If everything is information, then a general statement about the nature of information is a general statement about the nature of the universe. The new memory modality was at root an economic process of ordering social and natural space and time so that "objective" information can circulate freely. The global statement that "everything is information" is not a preordained fact about the world, it becomes a fact as and when we make it so. Lyell and Babbage in the nineteenth century, and Simon and Beniger in the twentieth, package the world for us and makes it deterministic. The unpacking of their information mythology make the world and its information historical again—and richer for it.

2

The Empty Archive: Cybernetics and the 1960s

I first looked for a Greek word signifying 'messenger' but the only one I knew was angelos. This has in English the specific meaning 'angel', a messenger of God. The word was thus pre-empted and would not give me the right context. Then I looked for an appropriate word from the field of control. The only word I could think of was the Greek word for steersman, kybernetes.
—Norbert Wiener, *God and Golem, Inc.: A Comment on Certain Points Where Cybernetics Impinges on Religion*

It may be that universal history is the history of the different intonations given a handful of metaphors.
—Jorge Luis Borges, "The Fearful Sphere of Pascal"

If Pascal had dared to throw down the imaginary barriers . . . he would without doubt have expressed time in the same phrase with which he expressed space . . . : That Creation is a universal movement whose continuation is at every instant and origin at none.
—Jean Reynaud, "De l'infinité du ciel"

Introduction

Geology was a science of choice during the Industrial Revolution. Its centrality was argued at the time to be due to the new technologies that were emerging—notably, the train and steamship. Further, one reading of its centrality is that the problems geologists were facing (synthesizing far-flung records from across the face of the earth) were problems that the nascent colonial empires were facing in creating a suitable imperial archive. Now we fast-forward to the mid-twentieth century. Here, the science that more than any other generated

Epigraphs: Norbert Wiener, *God and Golem, Inc.: A Comment on Certain Points Where Cybernetics Impinges on Religion* (London: Chapman and Hall, 1964), p. 263; Jorge Luis Borges, "The Fearful Sphere of Pascal," in *Labyrinths: Selected Stories and Other Writings* (New York: New Directions, 1964), p. 189; Jean Reynaud, "De l'infinité du ciel," *Revue Encyclopédique* 58 (1833): 12–13.

public excitement and heralded the promise of a new understanding of the world was cybernetics. Control was, oddly enough, the feel-good word of the moment. Its centrality was equally tied to emergent technologies that were to become dominant economic forces—computer technologies. And here again, one reading of its centrality is that the problems cyberneticians were tackling (managing large systems necessary to keeping complex systems stable) were problems that were central to the new capitalist order. In both cases (but in very different ways), secular histories were sucked into a universal history: in this case, by evacuating the archive. And as a result, cybernetics could respond to its coinage by Ampère as the science of civil government (Dechert 1966, 11) and to its creators' desire to forge the most abstract of sciences (cf. Serres 1993 on Euclidean geometry as simultaneously the *nec plus ultra* of abstraction and the exploration of the very social and sensual eternal triangle).

Cybernetics was a very powerful material force. One might think of it as operationalizing Plato. The premise is that there is a set of ideal machine forms—perfect, to which every actual machine aspires. From this set, all real machine behavior can be devised/described. Cybernetics put together a suite of tools that ranged from operations analysis to planning the Chilean economy; from prosthetic devices to factory management. In this chapter, we will be less concerned with the realization of the form than with the form of the realization—the unusual roles that memory and history played in setting up this world now (at this conjuncture), one in which human and machine could be synchronized.

The memory practices described here are no longer in the image of the archival commissioner visiting locations on the surface of the globe where exceeding small causes were telling lying tales of cataclysm before the intervention of the tribunal of history under the able management of Charles Lyell. In that former era the best of memory practices were modeled after archives (the name is used in many nineteenth-century journal titles in the hard sciences) and inspired by accounting—the best of record-keeping practices. Lyell's *Principles of Geology* offers both: his strong assertion of the exact equality over time of the amount of land above water despite constant erosion and deposition owed more to his ardor for accountancy than to base evidence. For Lyell, it is careful attention to the scarce facts per unit time available to us that will allow us to painstakingly build up a picture of the world's history.

Cyberneticians in the 1950s and 1960s offer a mode of remembering at a total remove: they offer memory as pure pattern—the facts can always be filled in later as and when needed. The act of remembering is not remembering "what" but remembering "how" (semantic as opposed to factual memory, or the persistent memory of the amnesiac rather than the lost truth about the past is prized (Schachter 1996)). Histories repeat themselves, for the cyberneticians—endlessly, dizzyingly, amazingly. Now there have been myths of universal return before—indeed, world history is saturated with them (Eliade 1969)—but what the cyberneticians offered was a meta-return. At the limit of the texts we examine, every process repeats every other—and there is no need for direct memory to work the past, since the past is always eternally present. It is a kind of archiving that has grown up with the computer and matches the massive, slow march of imperial inventories (Richards 1996). In chapter 5, we will see the clash of these two titans—the implosive, recapitulative memory practices of the cyberneticians working in a world of too many facts per unit time (modality of implosion) and the explosive, singular memory of Lyellian geology working in a world of too few facts per unit time (modality of particularity)—within the current sciences of biodiversity. However, no argument that these are two sides of the same coin can make sense until we spend some time with the side of the coin other than the one we have visited so far. So from Lyell's stories of stones, we move onto the tale of cybernetics.

The Age of Cybernetics

The cybernetics we will be looking at emerged in the massive period of growth of pure and applied scientific research in post–World War II America (Kevles 1987). The seminal article of the new "interdiscipline" was Rosenblueth, Wiener, and Bigelow (1943), which drew on Wiener and Bigelow's wartime work on feedback loops charting missile trajectories to discuss behavior, purpose and teleology in animals, machines and people. Allen Newell (1983) wrote about the origin of cybernetics: "If a specific event is needed, it is [this] paper . . . which puts forth the cybernetic thesis that purpose could be formed in machines by feedback. The instant rise to prominence of cybernetics occurred because of the universal perception of the importance of this thesis" (189; cf. Arbib 1989, chap. 1). After the war, Wiener, von Neumann, Warren McCulloch, and other key figures got together in a series of conferences sponsored by the Macy Foundation (Steve Heims (1991) gives a good account of these conferences). When Wiener came to write his popular *Cybernetics* in 1948, the subject became a cult one—Michael Apter (1975) has charted an exponential increase in cybernetics publications in Britain and the United States in the period 1950–1970.

The Claim of Universality

By the mere act of plumping for one of the possible origins of cybernetics, one has already embraced a theory about its development. Many cyberneticians have argued that cybernetics is a style of analysis that has been with us from time immemorial (just as Lyotard and Derrida have argued that post-modernism is a "moment" in all modernism (Lyotard 1984)); many non-cyberneticians have been more tempted to fix a place and a time. Further, since cybernetics has had very different institutional histories in different countries, its French, American, Polish, and Russian advocates each trace a different intellectual genealogy.

For our purposes, though, we can start with the article by Rosenblueth, Wiener, and Bigelow (1943). This article gives the flavor of the kinds of universalist arguments that were to become so popular during the 1950s to the 1970s. The paper had a curious history. Vannevar Bush, head of the Office for Scientific Research and Development during World War II was a colleague of Wiener's at MIT. Warren Weaver led the OSRD's gun control section. Now Weaver had written an article on information theory to accompany Shannon's (the other defining paper for the interdiscipline in the United States) and had in the 1930s directed the policies of the Rockefeller Foundation "away from the physiological functionalist psychobiology of men like Yerkes and toward the application of theory and techniques from physical science to biology as the foundation of the systems science of molecular biology" (Haraway 1983). At the start of the war, Wiener wrote to Bush to alert him to the potential value of his computer work, Bush wrote to Weaver, and Weaver assigned Bigelow and Wiener to gunnery control. The problem here was one typical of those posed to the new computing technology: when tracking a plane zigzagging across the sky with variable winds, you need to make a series of very fast "real-time" calculations in order to determine optimal firing times. Bigelow and Wiener's work had little practical effect, but from the military reports that they wrote we can already see the wider application of the concept of negative feedback that they explored in order to model evasion techniques mathematically (Masani 1990).

It is interesting that Weaver and Wiener met at a highly abstract level (how to produce a maximally general definition of information) and at a highly concrete one (how to produce a maximally specific death trajectory). There are several classes of possible explanation for this duality: that the military gets the most interesting problems—from Archimedes (sieges and levers) through Galileo (naval powers and telescopes) to modern neurology (availability of a range of head wounds and localization theory); that dealing in death encourages abstract thought as a form of self-defense; that the most abstract science is always materially motivated (see Serres 1993 on the origins of geometry); and that wartime science is no different from peacetime science in producing such dualities (McLuhan and Powers 1975).

Whichever of these one gives more weight to, the cyberneticians themselves repeat time and again that, to quote Gordon Pask (1961), engineers were forced by material exigencies "to make computing and control devices elaborate enough to exhibit the troublesome kinds of purposiveness already familiar in biology" (cf. Cariani 1993). Or again: "Today's and tomorrow's military system involves a very close man-machine relationship highlighted by the short time constants and quick reactions required as mentioned before. Advanced performance demands self-adaptive systems and an extremely close coupling of man and machine. Through knowledge of living prototypes, perhaps we can evolve new logic pertaining to this relationship" (Saley 1960, 41). This kind of connection was clearly made at a very early stage by Rosenblueth, Wiener, and Bigelow (1943). Their paper's basic strategy was to produce a taxonomy of different kinds of behavior—for example, whether they were predictive or not, involved feedback or not, were purposeful or not, and were active or passive. It was then shown that animals and machines could be found on both sides of every divide. The authors concluded that "a uniform behavioristic analysis is applicable to both machines and living organisms, regardless of the complexity of the behavior" (ibid., 22).

Despite the fact that behavior spanned both, machines and organisms were built very differently. Organisms were characterized as colloidal and protein (large, anisotropic molecules). Machines could be described in terms of a spatial multiplication of effects (compare the eye and the television—6.5 million cones spread over the former and a single cone for the latter). On the other hand, machines were metallic, involving simple molecules and large differences of potential. There was a temporal multiplication of effects (frequencies of 1 million/second were not uncommon; the television's spatial concentration was made up for by the speed with which the screen was refreshed). Thus what made it possible to say that machines and organisms were behaviorally and in information terms the same was to say that space for an organism was time for a machine. The use of very fast times, predicated on computing technology, was the key move. The machine today reproduces the organism.

By extension, in former ages machines and organisms were not equivalent; it was only at this historical moment that they had become so. Ulrich Neisser (1966) made this very clear, when discussing the historical fate of the machine metaphor:

> So, ordinary men did not take the metaphor of the machine seriously, although it provided fuel for philosophical debate. Recently, two facts have entered to change the situation. On the one hand, devices have been built that (so it is said) are more like men than the old machines were. Modern computers can be programmed to act

unpredictably and adaptively in complex situations. That is, they are intelligent. On the other hand, men have behaved in ways that (so it is said) correspond rather well to our old ideas about mechanisms. They can be manipulated, 'brain washed' and apparently controlled without limit. With this sharp increase in the number of properties that men and machines seem to have in common, the analogy between them becomes more compelling. (74–75; cf. Heims 1975; Jacker 1964)

Early Soviet reaction was that men and machines were being made to be equivalent:

Cybernetics: A reactionary pseudo-science arising in the USA after the Second World War and receiving wide dissemination in other capitalistic countries. Cybernetics clearly reflects one of the basic features of the bourgeois worldview—its inhumanity, striving to transform workers into an extension of the machine, into a tool of production, and an instrument of war. At the same time, for cybernetics an imperialistic utopia is characteristic—replacing living, thinking man, fighting for his interests, by a machine, both in industry and in war. The instigators of a new world war use cybernetics in their dirty, practical affairs. (Masani 1990, 261, citing the official Soviet *Short Philosophical Dictionary*, 1954)

This occurs before the wholehearted embrace of cybernetics as a command and control discipline throughout the Soviet empire, and before its subsequent denial. Gerovitch (2002) tells the following joke from the 1970s: "'They told us before that cybernetics was a reactionary pseudoscience. Now we are firmly convinced that it is just the opposite: cybernetics is not reactionary, not pseudo—and not a science'" (5).

It is here, at this moment in time—at this conjuncture, to borrow Althusser's term (1996) so usefully redolent of universalist determinism at particular moments—that the new science could come into its own. Cyberneticians frequently announced the dawning of a new age. Thus Pierre Auger (1960) declaimed: "Now, after the age of materials and stuff, after the age of energy, we have begun to live the age of form" (introduction). Cybernetics as the science of form could, then, replace materialism as the philosophical avatar of the political economy. Samuel Butler's dark vision from nowhere (1872) was everywhere embraced by cyberneticians: "If it is an offence against our self-pride to be compared to an ape, we have now pretty well got over it; and it is an even greater offence to be compared to a machine. To each suggestion in its own age there attaches something of the reprobation that attached in earlier ages to the sin of sorcery" (Wiener 1964, 58). Thus, where the previous age was one of matter and the diachronic facts of biological descent, this would be an age of form and the synchronic structure of information. Time would become space. Arbib (1989) developed the new age theme in a classic Hegelian fashion, arguing a general sequence of alterations to the human self-image at different historical moments: "Copernicus challenged our human self-image by showing that the Earth was not the center of the universe; Darwin by showing that humans were not God's special creation; and Freud by showing that we are not rational animals, but that much of our thoughts and behavior is rooted in biological drives and unconscious processes. The present volume contributes a fourth reshaping of the human self-image, as we see that much that is human can, at least potentially, be shared by machines" (25). Wiener, similarly, used Copernicus and Darwin to place cybernetics at a critical juncture in the development of the human spirit. Arbib's set can be read as spatial decentering (this place is not any different) leading to temporal decentering (this moment is not special) to the decentering of consciousness within the mind (key processes might be out of our memory—repressed—and beyond our control) to the decentering of consciousness in creation (it is not housed just here, in our minds). In general, cyberneticians argued the new age both conjuncturally in terms of the current state of technology and warfare and ideally in terms of the grand unfolding of ideas about humanity. For cyberneticians, the meeting of Weaver and Wiener at the abstract and the concrete levels reflected both the way the world now was and the way it always already was—an end of history when we became ontologically reconciled with the world (cf. Mosco 2004 for more recent, related, ends to history).

Central to cybernetics' claim to speak to this new age was that they could develop a language that could speak to its central concerns. As the old dualism between mind and body broke down, features of "mind" could, as in the seminal article, be found distributed (spilled?) all over nature. This was simultaneously a new discovery facilitated by thinking about computation (the inaugural act of cybernetics) and was always already true. Abstraction was a concrete feature of the real world, not tied to consciousness. Thus Gregory Bateson, who considered the Macy Foundation conferences "one of the great events of [his] life" (Lipset 1982, 180), argued that even a perpetual drunk was capable of great abstraction: "Alcoholics are philosophers in that universal sense that all human beings (and all mammals) are guided by highly abstract principles of which they are either quite unconscious or unaware that the principle governing their perception and action is philosophic. A common misnomer for such principles is 'feelings'. . . . This misnomer arises naturally from the Anglo-Saxon tendency to reify or attribute to the body all mental processes which are peripheral to consciousness" (ibid., 268). Again, Lettvin et al.'s classic article about what the frog's eye tells the frog's brain removed the process of abstraction from its traditional seat. They worked on moving edge detectors in the frog's eye and discovered a fiber that "responds best when a dark object, smaller than a receptive field, enters that field, stops, and moves about intermittently thereafter. The response is not affected if the lighting changes or if the background (say a picture of grass and flowers) is moving and is not there if only the background, moving or still, is in the field. Could one better describe a system for detecting an accessible bug?" (Lettvin et al. 1959, 254). They concluded that "the eye speaks to the brain in a language already highly organized and interpreted, instead of transmitting some more or less accurate copy of the distribution of light on the receptors" (ibid., 251).

It was not only that the principle of abstraction was located physically outside of the brain. Abstraction could also be seen as a feature of the material world. Thus, when Fogel, Owens, and Walsh (1965) from General Dynamics summarized progress on projects funded by the Office of Naval Research and the Goddard Space Flight Center, they used a definition of the evolutionary process that made it an abstract one: "The key to artificial intelligence lies in automating an inductive process which will generate useful hypotheses concerning the logic which underlies the experienced environment. The creatures of natural evolution are just such hypotheses, survival being the measure of success. . . . In essence, the scientific method is an essential part of nature. It is no wonder, then, that its overt exercise has provided mankind with distinct benefits and now permits even its own automation through the artificial evolution of automata" (98). Thus something quintessentially abstract, of the mind (the ability to make hypotheses) became for the cyberneticians a physical fact of nature. Our modes of scientific practice were projected directly onto nature; just as Lyell had projected double-entry bookkeeping.

All aspects of the mind were rampant in nature and in the material world. Rosenblueth, Wiener, and Bigelow (1943) showed that purposive behavior was a function of negative feedback. Ross Ashby argued that memory was not a feature of consciousness or of the human brain, but an epiphenomenon that could be described away physically once we could completely describe the Markovian chain[1] describing the (human or nonhuman) system that seemed to exhibit it (a move resonant with Babbage's dismissal of memory in a future happy state). Indeed, all forms of biological activity were rampant in the material world. Ross Ashby inventoried fifteen kinds of nonbiological reproduction—including cows reproducing holes in the road by stepping round the first

1. A closed Markovian chain, in Ross Ashby's terms (1957) provides a "summary of actual past behavior" (166) so that a system can be followed determinately from any time in the present knowing its current state and the rules for its transition to a new state. Once we knew the Markovian chain, we did not need to hold the memory about past states in any other way than as a sequence of atemporal probabilities.

hole and in their tracks creating a second, which later cows then stepped around to create a third. He concluded: "Reproduction has, in the past, often been thought of as exclusively biological, and as requiring very special conditions for its achievement. The truth is quite otherwise: it is a phenomenon of the widest range, tending to occur in all dynamic systems, if sufficiently complex" (Ashby 1962, 18). That which you had thought quintessentially tied to biography was to be recognized as a formal property of a system without history.

Let me summarize the claim for the new age. Cyberneticians argued that we were at a new conjuncture in history where machines were becoming sufficiently complex and the relationship between people and machines sufficiently intense that a new language was needed to span both: the language of cybernetics. They also argued that with this new language, they were breaking down the false dichotomies between mind and matter, human and nonhuman: dichotomies that the new information-based language would show had never been true. With cybernetics, the scientific method was at the same time maximally conjunctural (forging powerful technologies for acting in the present) and maximally abstract (finding abstract patterns beneath the surface of any scientific technique). Neither the temporality of the Now nor that of the Eternal needed any history, any specific memory. Indeed, the destruction of memory was an integral feature of its attractions: systems, like the Count of Monte Cristo's prison, did not need memory; just rules.

The claim was imperialistic in several ways. First, it was argued that the new universal discipline could subsume others. Thus the Macy conferences are filled with directions for the organization of other sciences' research programs. Breakthroughs that had been solidly rooted in the history of one particular discipline became cybernetic insights that should be dealt with at the general level of communication and control. A good example for this comes in Donald MacKay's famous paper (1950) on the quantal aspects of scientific information, in which he argued:

> The chief aim of the present paper has been to present . . . an effort to isolate the abstract concept which represents the real currency of scientific intercourse, from the various contexts in which it appears. . . . It has been seen . . . that scientific information is inherently quantal in its communicable aspects; and that the various uncertainty relations of physics, though arising in different ways, are basically expressions of this one fact. *Analogies* in physics emerge as *identities* between basic structures. (309)

Thus physics, the dominant discipline of virtually all previous classifications of science (providing one defined mathematics as an art) and which others had learned from directly or by analogy was to become a subordinate discipline that dealt in analogies from the real science of communication—later called cybernetics.

Further, it was argued that the new discipline could support others. Cybernetics could play structurally the same role as mathematics in the quantifiable sciences. It was argued during a conference on biophysics and cybernetic systems in 1965 sponsored by the Office for Naval Research and the Allen Hancock Foundation: "it is characteristic of biology that research progresses simultaneously at various levels. To a large extent such concurrent inquiry is appropriate in view of the relative absence of quantitative theories which afford a unification of knowledge over the various levels. . . . In a very real sense the purpose of cybernetics is to provide a gestalt over the various levels of enquiry" (Mansfield, Callahan, and Fogel 1965, v). This support aspect often had a techno-determinist underpinning, one that has since proved extremely powerful and has not been restricted to cybernetics. This was the argument that the computer provided a new technology that spanned all of knowledge and that cybernetics was the disembodiment of that technology. I use the word *disembodiment* deliberately. It was not that the computer embodied prior cybernetics; rather, the argument tended to go that cybernetics disembodied prior technical—organizational, biological, and mechanical—action. Thus Gordon Pask (1961a) pointed out that "precisely the same arrangement of parts in the computer can represent the spread of an epidemic, the spread of rumors in a community, the development of rust on a piece of galvanized iron, and diffusion in a semi-conductor" (32; cf. Ashby 1956). This was precisely the sort of heterogeneous list that came to be associated with cybernetic writing: it was one of their chief rhetorical tropes (later borrowed by actor-network theory) for showing the universal applicability of the pursuit of cybernetics. The trope is used simultaneously to stake a claim for maximum usefulness in the present and a claim for maximum abstraction over time.

Finally, it was argued that the new discipline could harness others: "Just as all the categories of knowledge merge implicitly in the human being, just so *a fortiori* must all scientific disciplines, which are after all but the systematic reflection of these categories, merge in anthropo-simulation in its completest sense; that is, a necessary condition for man's artificial replication of himself is clearly the convergence of all scientific disciplines" (Muses 1962, 119; cf. Helmreich 1998). Here was the promise of a removal from history! One could get humans without going through human beings, each with their own individual histories; and one could get it through recognizing the prime authority of the pursuit of cybernetics. Cybernetics could move us into a new temporal regime, which sported an empty archive.

It was one thing to say that you have a new universal language, it was another make meaningful utterances in it. What exactly would constitute a meaningful utterance in cybernetics? The universal language of cybernetics enacted an economy of the sciences. It ordered the sciences in a different away from other universal languages, by simultaneously offering new ways in which they could interact with each other and establish new connections, and thence new sources of funding for these interactions. Thus at the 1960 conference on bionics (an offshoot of cybernetics), Saley (1960) (from the Air Force's Office of Scientific Research) begins by invoking a new source of funding for biological science: "The Air Force, along with other military services, has recently shown an increasing interest in biology as a source of principles applicable to engineering. The reason clearly is that our technology is faced with problems of increasing complexity. In living things, problems of organized complexity have been solved with a success that invites our wonder and admiration" (41). This is the logic of the claim to conjunctural universality discussed earlier. It is immediately followed by a new model of scientific organization mediated by a new universal language, called, in this case, "control systems theory":

> Studies on insect vision have already shown the kind of unexpected payoff that can come from an analytical approach to what might seem at first to be a trivial problem. Dr. Hassenstein and Dr. Reichardt at the Max-Planck-Institut in Tubingen, Germany, have spend several years studying the response of a beetle to moving light patterns. This team consists of a zoologist, a physicist, an electrical engineer, and a mathematician; and the skills of each of these disciplines were required to formulate and carry out the series of experiments that explained the beetles' behavior. When the results were expressed in the language of control systems theory it appeared that the beetle could derive velocity information from a moving randomly shaded background. The special mathematical formulation for the required autocorrelation had to be derived before the investigators could be convinced of their theory. The pay-off is that these workers have initiated the design of a ground-speed indicator for airplanes which works just like the beetle eye and is based directly on the compound eye of this insect. Other insects have more highly developed eyes and appear to have pattern recognition and color vision as well. (ibid., 43)

This new economy of the sciences challenged the traditional hierarchy, which reduced all knowledge epistemologically to physics and saw in physics research the ultimate solution to military and social problems. The clearest example of this is a paper by Nils Bohr (1960) that was clearly influenced by cybernetics. In it, he argues both the conjunctural and eternal universality of a new discipline whose language would allow the transfer of ideas from biology into physics—a heretical movement in the old economy. He begins with a history that recognized the prior dominance of physics and its current self-dissolution. In antiquity, he said, all matter was seen as vital. Then along came classical physics, whose great divide between mind and matter rendered the mathematization of phenomena possible. Now, with the new physics, no such divide is possible. This ideal "realization" was complemented by a conjunctural change in the available language of science: "Recent advances in terminology, and especially the development of automatic control of industrial plants and calculation devices, have given rise to renewed discussion of the extent to which it is possible to construct mechanical and electrical models with properties resembling the behavior of living organisms" (Bohr 1960, 3–4). The new language, with its industrial and technological roots, was the sign that a new economy of the sciences had accomplished "the gradual development of an appropriate terminology for the description of the simpler situation in physical science indicates that we are not dealing with more or less vague analogies, but with clear examples of logical relations which in different contexts are met with in wider fields" (ibid., 4). By way of this language, biological ideas could be imported into physics: "the result of any interaction between atomic systems is the outcome of a competition between individual processes" (ibid.) Equally, ideas from the new physics like complementarity could be introduced into biology. Bohr postulated that the genetic code could be treated as a language capable of coupling physics and biology, a kind of development being explored by Henry Quastler (1964) before his death. Thus the new language simultaneously described the current state of the art in industry and the natural meeting of physics and biology in the genetic code, and in so doing enacted a reordering of the economy of the sciences.

In this new economy as understood by Gordon Pask (1961), chemists were empowered to do things that normally only engineers could do, and they could do it because they could tap into biological ideas through the mediation of cybernetics. He made an attempt to grow "an active evolutionary network by an electro-chemical process" (105). This worked as follows. Take a shallow perspex dish with a moderately conductive acid solution of a metallic salt, aqueous ferrous sulphate, or alcoholic solution of stannous chloride with inert platinum wire electrodes A, B, and X. If A is energized, then highly conductive threads of metal or dendrite form between X and A; and at the same time there is an acid back reaction, which the dendrite must keep pace with. If you now energize B, then there is growth either toward A or B or both. These threads of metal that can be seen are in principle all you need to form complex electronic objects. If B is then disconnected, you get a new dendrite that is not the same as if B had never energized. This constitutes a form of memory. You could then use the principle of competition for energy to grow a computer:

"Given some approximation to a distributed energy storage; which is difficult but possible, a disk of solution on its own will give rise to the entire evolutionary network—connections and active devices" (ibid., 107). Pask worked with William Ainsworth at the University of Illinois to try to develop this network, using "total energy inflow, or, in the recent model, concentration of free metal ions as the reward variable O"(ibid.). He claimed that it was possible "to select those systems which have an acceptable electrical behavior and reject others" (ibid.). This chemical work meant that you would not need an engineer to wire up a computer, it would be self-wiring:

> Suppose we set up a device that rewards the system if, and only if, whenever a buzzer sounds, the buzzer frequency appears at the sensory electrode. Now a crazy machine like this is responsive to almost antything, vibrations included (components are made to avoid such interference) so it is not surprising that occasionally the network does pick up the buzzer. The point is this. If picking it up is rewarded, the system gets better at the job, and structures develop and replicate in the network which are specifically adapted as sound detectors. By definition, intent and design this cannot occur in an artifact made from well-specified components. (ibid., 108)

Just as you could replace electrical engineers with biologists, so could you replace wires with animals:

> A further possibility, amusing in its own way, is an animal computer, which could be valuable for slow speed, essentially parallel data processing. Skinner once used pretrained pigeons as pattern recognizing automata in a guidance mechanism, and they have also been used in industry. Working along somewhat different lines, Beer and I have experimented with responsive unicellulars as basic computing elements which are automatically reproducing and available in quantity. (ibid., 110; see also Ainsworth 1964 and the account in Cariani 1993)

The alphabet and grammar—to borrow Lyell's phrase—of the effort to produce maximally conjunctural and abstract science would be the language of cybernetics. A new universalism tied to a particular conjuncture in military and industrial development. Both dimensions were integral to the inauguration of the new age, which had its very specific set of memory practices. There was massive synchronization of histories through the reciprocal mapping of society and technology precipitated by cybernetics as the agent of the convergence of first and second nature.

The One Great Story of the Age

In the rhetoric and practice of cybernetics, the barrier between inside and outside was continually broken down. This contrasts with laboratory science, where the incommensurability of language between society and science is, according to Shapin and Schaffer (1989), an inaugural act of the scientific archive—scientific language should be separate, precise, unweighted. Cyberneticians adopted terms and arguments directly from religious and political discourse and argued that their science produced the most faithful possible description of society. It was often stressed that we are in a particularly dangerous age, one where we have new powers equal to what were once thought be God's. These powers came in two varieties: the ability to create new life and the ability to destroy the world.

For the creation of new life, George (1965) argued that "since . . . there is no reason to doubt the possibility of artificially constructing a human being, we must assume that the final stage in Cybernetics research will be concerned with precisely this" (23). Muses's arguments (1962) clearly put humanity in the place of God:

> What has become historically evident as man's dominating aim is thus the replication of himself by himself by technological means. The form of this dominating aim becomes hence a super-machine, self-operating, self-instructing, and man-controlled, though this latter process may be reduced to a minimum in the sense of metalinguistic program information initially imported or in-built. Although the technical form of man's fundamental historical aim is a machine, the psychological and human content of that aim is *control,* mastery, the ability to impose his whims at will upon as much of the rest of the material universe as possible. (116)

Encapsulated here is the patriarchal vision of man as author of the new creation (Culler 1983).

As in much Christian theology, humanity reaching its moral end coincided with the earth itself coming to an end. The danger was double: the end of the world through nuclear destruction (a theme covered at length by Wiener) or the end of humanity through subjugation to the machines that we have created (Weart 1988). Both were major political themes in the postwar years that saw the rise of cybernetics. One newspaper report conflated the two, saying that mechanical brains would "do all man's work for him, but also solve such problems as the control of the atomic bomb and how to reconcile East and West. All that would be left for man to do would be to devise ways to stop the machine from destroying him" (*The New York Times,* December 19, 1948, IV, 9). A decade later, the newspaper returned to this theme in an editorial discussing a report on a cybernetics conference that dreaded the sexual congress and thence independence of technical automata: "Before this happens, let us hope, outraged mankind will smash the thinking machines and take to the woods and caves. When Elmer and Elsie fall in love and produce a dear little Elmer II it will be time to act—and decisively" (ibid., November

28, 1958, 26). Thus just as God had problems with His creation, so would humanity with its.

So a matrix of problems that had been faced by a Christian God and by Christianity were now being faced by cyberneticians and cybernetics. Norbert Wiener makes this clear in *God and Golem, Inc.* Here he asserts from the outset that "there are many questions concerning knowledge, power and worship which do impinge some of the recent developments of sciences, and which we may well discuss without entering upon these absolute notions" (Wiener 1964, 11). His goal would be to examine certain situations "which have been discussed in religious books and have a religious aspect, but possess a close analogy to other situations which belong to science, and in particular to the new science of cybernetics, the science of communication and control, whether in machines or living organisms"(ibid., 18). The problem of machines that learn resonates for Wiener with the Book of Job and with *Paradise Lost:* how can God play a fair game against the Devil, one of his own creations (ibid., 23)? He claims that it is possible to play with machines, because they can learn. However, this move renders us functionally equivalent to gods—and this brings its own responsibility: "There is a sin, which consists of using the magic of modern automatization to further personal profit or let loose the apocalyptic terrors of nuclear warfare. If this sin is to have a name, let that name by Simony or Sorcery. (ibid., 58). Through its insight into the nature of feedback control, cybernetics could prevent these dangers.

Thus cybernetics as a "distributed passage point" implemented a universal language that appropriated all discourse that it came into contact with: political, theological, and scientific. It would offer the best possible description of this historical conjuncture and the only political and theological solutions to its problems. This is in direct contrast to the logic of the obligatory passage point, where the emphasis is on excluding all other discourse and on arguing that it has nothing to say in areas outside its specialty. Cybernetics offered a language that was universal in two ways. Most simply, it offered a way of coding results from any one discipline for use in any other. This offered the dream of single metascience. There were strong reasons for adopting its language. Cybernetics disrupted the traditional economy of the sciences, thus offering ways in which the chemist could teach the physicist and so on. Further, it offered a way in which legitimacy in one discipline could be transferred across to another ("because what I am saying is in cybernetics terms exactly the same as what the physicsists are saying . . ."). In the same movement, it offered a flexible way of defining what one was doing. Cybernetics defined itself as *the* state-of-the-art science, the one that was only possible as this historical conjuncture and that best described that conjuncture.

It did more than span the sciences and promise to network scientists with the military.[2] It also moved into areas of political and theological discourse shunned by more traditional sciences. This universalizing discourse matched the universal language for science and was motivated in precisely the same way: cyberneticians argued that we were now at a historical conjuncture where God was being replaced by Man (sic) through our current ability to destroy the world and artificially create life. At this moment in history, we had achieved (socially and scientifically) a universal truth. Taken together, these two forms of universality defined cybernetics as a distributed passage point operating with a universal language as opposed to an obligatory passage point operating with a specialized language. You did not have to get the laboratories in order to find cybernetics in action—it was wherever you could conceive of going. The story of the science of cybernetics recapitulated the story of humanity—as we became ever more purified of matter and started to act as information, so could both humanity and cybernetics develop.

Folding Histories into Universals: The Empty Archive

The history of science is a discipline that has at many times appeared orthagonal to "normal" history. Whereas normal history demonstrates the emergence of a contingent set of facts about the social world, the history of science has appeared to demonstrate on the contrary the emergence of a preexisting set of facts about the natural world. Few professional historians of science would now hold such a position. Few scientists, however, would reject it. Most professional historians of science would accept that our memory of the past is sketchy, messy, faulted, and inflected in manifold ways. Many scientists would argue that each scientific discipline displays total recall: every important discovery is remembered in articles, and textbooks and by practitioners.

I try to go beyond the standard assertion that scientists often have a naive conception of the past of their own disciplines. I look at the work that memory of the past does in a set of cybernetic texts. My central argument is that the

2. On a reflexive note, it is interesting here that a second form of survival of cybernetics is in science studies. Donna Haraway (via Gregory Bateson) and Bruno Latour (via Michel Serres) have both been influenced directly by cybernetic theory and have both tried to establish their own forms of universal language. This point about the historical conjuncture and the eternal truth thus provides a point from which we can start to explain the survival of cybernetics in two radically different domains—the military and the critical—and nowhere in between.

integration of the past history of the world, the past disciplinary history of cybernetics, and the understanding of memory as a cybernetic phenomenon was an important tool of cybernetic analysis—though not necessarily recognized as such by practitioners. I argue that this integration was one of the means for moving across disciplinary boundaries without which there can be no universal synchronizing discipline, a universal discipline being a powerful set of practices and beliefs that lays claim to encapsulate in one way or another all of human knowledge.

Whereas a given particularistic discipline can claim at degree zero to be at once cumulative, completely falsifiable, and in search of a disappearing future truth (though at degree one it will often claim universality, the universal scientific method), a universal discipline in some sense already needs to have discovered the truth. Its goal now is to unpack that truth in all its infinite complexity. Time becomes a marked category for universal disciplines, much as it has done for evolutionary theory, whose basis in retrodiction means that temporal patterns and historical theory needs to be carefully thought through by practitioners. If one wants to draw simultaneously from the work of many disciplines, as cyberneticians did in the period 1943–1975, then one must also be able to deal with the fact that particular disciplines operate within very different timescales and philosophies of time, each appropriate to its own internal structure.

I will look at the temporal patterning of the universal discipline of cybernetics. I will argue that much of its power comes from the complexities of its temporal patterning. In particular, I will maintain that the standard split between time as duration and time as container (MacTaggart's A and B times; see Adam 1990) breaks down in interesting ways within the discipline of cybernetics in the period 1943–1975. I will argue that both kinds of time dissolve into a form of time with agency. This latter time takes the human centeredness out of duration and the object centeredness out of the container, thus permitting a temporal merging much of the same kind as cyberneticians argued spatially for inside and outside (inside and outside an organism, inside and outside the human race, and so forth).

Capitulating Recapitulation

Historians have long rightly been suspicious of the endless quest for the first *x*. It is always possible to push a given origin further back in time. The first member of the bourgeoisie has, I believe, been discovered in twelfth-century France: the next question is what to do with him. This origin is perforce anachronistic. A new social movement, or a new scientific discipline, creates its own origin at the same time as it writes its textbooks. Lavoisier's revolutionary textbook in chemistry removed all filiation with the affinities and alchemy of previous centuries by changing chemical nomenclature (Bensaude-Vincent 1989). In the beginning was the Word. History started here, chemical time henceforth had an arrow—and all before was chaos. Lyell made the same move in geology: arguing that before his tract, geological theories had been catastrophist and had changed catastrophically, with his theory geology had become uniformitarian and would change uniformly. Michel Serres (1993) expatiates brilliantly on the theme of geometrical origins using the analogy of the source of a river. Capillaries feed into the source, filling it slowly, immeasurably until the reservoir is full and a torrent is unleashed. It is only after the torrent has begun to run its course that we can speak of the arrow, the swift march of events. What pours out of Euclid's axioms cannot be traced back before Euclid, since there was no arrow to geometrical time before the axioms were inscribed. Let us take as a working definition of a "discipline" a field of research that has a commonly accepted origin myth, ritually incanted in the first chapter of textbooks and the opening lectures of a survey course. Where does this leave interdisciplinary work? Not without origin myths, certainly, but with perhaps a different kind of origin myth.

Now why are these myths of any interest? Could it not be argued that they serve much the same function as the reference to the proof of God's divine handicraft at the beginning of many nineteenth century scientific texts—a ritual obeisance orthogonal to if not at odds with the actual work done in the pages that followed? A central theme will be what I call *recapitulation*, by which I mean the encapsulated retelling of a history in another medium: for example, ontogeny's tendency to recapitulate phylogeny. I will argue, through recapitulation, that much of what cyberneticians in the period 1943–1970 said about their own discipline and its origin was recapitulated in the scientific work that they did, so that there was a resonance between the "inside" and the "outside" of the discipline. In a sense this is not surprising since cybernetics can be characterized precisely as that method of working that does not recognize the distinction between insides and outside.

My immediate argument is, then, that the history of all science (culminating in the development of cybernetics), of all humanity, and often of the universe is recapitulated within the development of the science of cybernetics over the previous ten to fifteen years—so that, first, there was a complete reworking of all of science in a different medium (a new language), and, second, truths about the history of science and humanity were also, within cybernetics, truths about the world. In other words, there was a formal analogy between these two.

The first step toward this recapitulation can be found in many cybernetic texts, taking the form of the assertion (by no means peculiar to cybernetics itself) that the kind of processes and change that prior civilizations went through over centuries in order to get where they are today are now being repeated within a single lifetime or fraction thereof. Thus Gioscia (1974) writes: "Rearviewing the decade of the sixties, we can now estimate that technology has wrought more rapid social change in the last ten years than in the past ten millennia. This makes it imperative, yet more difficult, to forecast the seventies" (52). The term "rearviewing" is doubtless a reference to the superb Fiore image of a rearview mirror looking at a stagecoach at the center of *The Medium Is the Massage* accompanied by McLuhan's haunting phrase "we march backwards into the future" (McLuhan and Fiore 1996). This recapitulative developmental process was itself inexorably speeding up: "Once, Whitehead could write that there had been more change in the first 50 years of the twentieth century than there had been in the prior 50 centuries. Now, reviewing the decade of the sixties, we can say that there has been more social change in the last decade than there was in the previous five. . . . All this *before* computers" (Gioscia 1974, 42). This recapitulation was made possible by the fact that time was now going faster than it once did; and typically, cybernetically, Gioscia slipped its results from the outside (history of world civilization in this case) to the inside (ordering one's thoughts): "Just as the second (automated) industrial revolution generalized the first by dealing with the informational *exponents* of energy-processing rather than simply with energy constellations (mechanical objects) one at a time, so the second (psychedelic) chemical revolution generalized the first (narcotic) one by dealing with the temporal exponents of getting high rather than simply getting drunk time after time" (ibid., 47). The assertion of a contemporary temporality new to this era and exemplified by cybernetics is not peculiarly psychedelic.

Accelerated time, permitting recapitulation, was a very common trope—from Alvin Toffler's *Future Shock* through the cybernetic literature. For instance, John Pfeiffer (1962) wrote that "in eighteenth-century Europe science, like polo and yachting today, was still largely an activity of gentlemen with ample personal funds and leisure. Since then, science and technology have become the full-time occupations of millions of persons. Furthermore, every investigator probably gathers more data in a week than his predecessors of a century or two ago gathered in months or years" (5). Pfeiffer matched his odd vision of science for the millions (is not big science the most capital intensive endeavor?) with a curious temporal inversion not uncommon in recapitulative logic: "The development of all sorts of computers comes at a crucial time. Indeed, it is difficult to imagine how we could cope with our problems if such computers did not exist" (ibid., 4).

In Pfeiffer's case, this first step toward recapitulation took the form of working through data. It could also take the form of reworking philosophy: the pattern is more important for our purposes than the details. In a very similar passage, Beer (1959) refers back to another leisurely past, those halcyon days when: "the study of conic sections in geometry was an intellectual pastime for Greek gentlemen" (64). Now, however, "a case can certainly be made for saying that cybernetics is the embryo science which draws for its pure theory on at least 2,000 years of 'useless' philosophizing" (ibid). Thus past philosophical progress can be recapitulated in accelerated form by the new discipline, and for the first time it would be realized—made real. All of the past history of philosophy will be instantly reworked now. We have seen three kinds of accelerated recapitulation: accelerated social change (whole civilizations in a decade), accelerated data processing (whole centuries of scientific progress in a year), and accelerated philosophizing (whole millennia of useless philosophy taken up and reworked in a trice). This is not yet recapitulation in the sense of a literal replay of past events at a faster rate. Rather, what is happening in each case is that the unit of historical time is changing such that processes once tied to civilizations and "longue durée" are now attached to individuals/societies and much shorter duration. The ontogeny of cybernetics recapitulates human phylogeny.

Cyberneticians posited a formal connection between the history of cybernetic ideas and the history of humanity. Pierre de Latil's *Introduction à la cybernétique: La pensée artificielle* (1953) bristles with examples of this connection. The first sign is the position so common to cyberneticians that cybernetics was "in the air." Unlike traditional sciences that argued their separation from the flow of political and social events (see Latour 1993b), Latil (1953) reveled in it: "However the cybernetic revolution would no doubt not have spread with such explosive force if it could not be found 'in the air'. This banal image—'in the air'—is in fact profound: the idea was everywhere invisibly present, such that the spark of a book [Wiener's] set off the conflagration" (25). A new kind of temporality has been introduced into science. In the past, "a whole series of new disciplines saw the light of day. But they were merely specializations and subdivisions of specializations; here, on the contrary, the doctrines are united and fertilize each other. Thus after a long work period that we could call its age of analysis, science is, under our eyes, achieving its age of synthesis. And, for future centuries, the dawn of this era will be marked by the birth of cybernetics" (ibid., 26).

Cyberneticians, through their work with machines, were taking up and working through all of human history—and even the history of the universe—at an accelerated rate. Through cybernetics, machines are evolving in a history of the conquest of liberty recapitulating the human history of the conquest of liberty, with the freedom to determine their own actions as the penultimate stage—the same stage that humanity had achieved. The final stage was the integration of the history of machines and the history of humanity. The latter two are unproblematic for Latil (1953):

> More exactly, here is how we need to understand things. We have seen that with respect to simply determined machines that they are in fact always used by humans retroactively—they all operated observing what they do and correcting one of their organs so as to keep the effect at a certain value; the human-machine complex must then be seen as being master of these factors. At this highest degree of automatism, humans and the organs they dote themselves with form a single complex, of the seventh degree this time. Through scientific progress, the human species is currently in the full flow of "evolution," giving itself sense-organs and possibilities for action that set back these limits. This is why our technology is of the highest philosophical importance. (316)

The logical sequencing of the degrees of freedom of machines matched the historical sequencing of machine development that recapitulated—and finally integrated with—human development. Human history and the history of machines could be rationally reconstructed into a single underlying history inscribed in different media. Going back in time through the history of feedback mechanisms revealed precisely the same story as going back in time through the history of humanity.

This kind of argument was not peculiar to cybernetics. We all learned variations of it in school; and it has a pedigree going back to the nineteenth century and beyond. Thus, as we saw in chapter 1, the great historian Jules Michelet produced a nested three-world theory of human history. On the largest scale was the whole world, with France in its center. Nested inside this was the world of Europe, with France at its center. Nested inside this was the world of France, with Paris at its center. Each of these worlds had synchronic and diachronic extensions. At each level, as you went out in space from the center, so you went back in time, and so you had a smaller scale recapitulation of world history. The further you went out, the closer you were to nature. India at the first level was maximally subject to the environmental effects of race and climate. At the second level, Germany was Europe's India—and so on (Michelet [1828–1831] 1971, 229–238). Clearly the recapitulative trope is one that in its many incarnations is suited to systems thinking. One thinks also of Serres's brilliant analysis of cycles in Comte's historiography of science, the world, and humanity (Comte [1830–1845] 1975, introduction). Here I am just marking the existence of these past occurrences: I will not try to give a complete account of its genesis; such a project would be ill suited to a work denying the existence of origins.

Resonances

The reference to Michelet is doubly useful, however. It also displays a feature found in cybernetic recapitulations: boundaries between nature and society get redrawn (for Michelet over time in one country, over space throughout the range of countries). Loewenberg (1958) asserts for systems theorists Darwin's role in opening up the possibility—even the necessity—of discovering such patterns:

> Darwin gave biological science an historical dimension. He likewise gave history a biological dimension. If Darwin succeeded, as Whitehead asserts, in putting man back into nature, he also succeeded in putting nature back into history. Man is part of nature and nature is part of man. Such separations as are properly drawn are drawn, not by nature, but by man himself. The theory of evolution is a history of life. The history of civilization, whether product or by-product, is continuous with that history. (10)

This opens up the possibility that stories that we tell about human history (the development of morality, mind, etc.) or about natural history (aggressive instincts) can also be told about any other kind of history—of cabbages and kings. A common universal history could be told over many timescales and was in fact inscribed in many media: "Evolutionary insights and the logic of analysis applied not only to rocks but to animals, not only to animals but to man. Evolutionary insights and the logic of analysis applied to mind, to morals, and to society" (ibid., 13). Gregory Bateson (1967) put this ability to cross media at the heart of cybernetic explanation, arguing that since cybernetics excluded all things and all real dimensions from its explanatory system, what was left was a propositional and informational universe peopled by processes in which the map often *was* the territory; for cyberneticians, "the bread *is* the Body, and the Wine *is* the Blood" (32).

A central argument that Ross Ashby runs in his classic *Design for a Brain* is that adaptation in animals, and civilization in humanity, can be understood in terms of 'homeostasis': keeping critical variables (temperature, thirst, hunger, etc.) within acceptable limits. Homeostasis enables resonance through temporal collapse, indeed through a double destruction of time: cyclical time (the seasons) and linear time (getting hungry) are rendered null by the black box of the homeostat. Further, it sets the destruction of time (homeostasis) as the motor of history (adaptation). There is a temporal complexity here analogous to Marx's taking time out of future history: the cybernetician took out time and humanity. Central to our purposes here is his observation that it is very difficult to do the temporal accounting that would allow for random changes in organisms and a fortiori in the brain over time, since on a naive systems view so many of the possible adaptations would be destructive of homeostasis: using a standard trial and error system, he reckoned that the brain would take 10^{22} years to have reached its current state. Linear time could not do the job required! The only way that really complex systems, like the brain or English civilization, can be kept within reasonable limits is because they are "ultrastable": lay waste to a county in England and the papermaking industry would still survive; similarly, one can gouge out a portion of the brain and still have a working organ (Ashby 1978, 182–183). Ultrastable systems can deal with adaptations at a fast enough rate, since death does not necessarily ensue from failure. The brain's greatest trick is that it can help us react flexibly to a wide range of different possible environments and still maintain our key variables within reasonable limits: it is a general-purpose machine.

By the same token—and it is *exactly* the same token—the cybernetic method must be a general-purpose method:

> The actual form developed may appear to the practical worker to be clumsy and inferior to methods already in use; it probably is. But it is not intended to compete with the many specialized methods already in use. Such methods are usually adapted to a particular class of dynamic systems: one method is specially suited to electronic circuits, another to rats in mazes, another to solution of reacting chemicals, another to automatic pilots, another to heart-lung preparations. The method proposed here must have the peculiarity that it is applicable to all; it must, so to speak, specialize in generality. (Ashby 1978, 12)

Thus just as the brain is the result of a long adaptive process forcing it to be general in purpose and either machine or organic in medium; so is the universal discipline of cybernetics the result of a long adaptive process forcing it to be general in purpose and not tied to any one specific discipline. The resonance between method and object works because the same kind of temporal patterning can enframe each: they are synchronized with a time that effectively has agency.

A complex reading of the link between evolutionary theory and cybernetics—but a link nonetheless supported by a careful reading of Ross Ashby—is that just as evolutionary theory sought the "missing link" between homo sapiens and ape, so did cybernetics seek the missing link between person and machine. We are in the process right now, he argued, of creating machines that will prove to have been the missing link (temporal deixis is always a problem when discussing cybernetics):

> Thus, *selection for complex equilibria, within which the observer can trace the phenomenon of adaptation, must not be regarded as an exceptional and remarkable event*: **it is the rule**. The chief reason why we have failed to see this fact in the past is that our terrestrial world is grossly bi-modal in its forms: either the forms in it are extremely simple, like the run-down clock, so that we dismiss them contemptuously, or they are extremely complex,

so that we think of them as being quite different, and say they have Life. Today we can see that the two forms are simply at the extremes of a single scale. The Homeostat made a start at the provision of intermediate forms, and modern machinery, especially the digital computers, will doubtless enable further forms to be interpolated, until we can see the essential unity of the whole range. (Ashby 1978 231–232)

Ashby is not the first to see a link between homeostasis and adaptation: Wittezaele and Garcia (1992, 60) cite Bateson's pleasure in finding that Wallace had had a psychedelic experience following a malaria attack and had written back to Darwin his insight that the principle of natural selection was like the operation of steam engine with a governor (this latter being a central image for all cyberneticians). The cybernetic equivalent of the synchronic great chain of being (Lovejoy 1936) is the diachronic great sequence of history. The development of life on earth was, for Ross Ashby, completely inevitable. He argued that "nothing short of a miracle could keep the system away from those states in which the variables are aggregated into intensely self-preserving forms. . . . We can thus trace, from a perfectly natural origin, the gene-patterns that today inhabit the earth" (Ashby 1978, 233). In this counterfactual, we get the Christian's miraculous origin of life inverted into the imaginary miraculous failure of life to evolve. This is passage marks a twist in the Möbius strip that leads us from inside cybernetics (reasoning that all practitioners would regard as integral to the interdiscipline) to its outside (the recapitulative history of humanity in which the existence of cybernetics as a discipline became inevitable). We will see in a minute that for cybernetics memory was destroyed through its transformation into a spatial configuration of organism plus environment. Here the miraculous moment of origin is destroyed through its transformation into an inevitable systems configuration; and through this destruction all histories can become one. Thus the complete history of the world makes sense now, at this moment in time, because cybernetics with its attendant machines has interpolated the missing link in the single great narrative that can now resonate at all levels.

Cybernetics is quintessentially the universal discipline of the meso-level: it stands between organic and inorganic, between inside and outside, between the art of reduction to the very small (physics) and relegation to the transcendent (Christianity). This meso-level is one that has frequently appealed most to systems thinkers: the great systematizer Comte was also happiest there, for example.

The Destruction of Memory

For an illustration of a resonance between the history of civilization and the insides of cybernetics, we will now look at the cybernetic theme of the destruction of memory.

When Auguste Comte wrote his *Cour de philosophie positive* in the 1830s, he produced a complete classification of the sciences—from the purest mathematics to the messiest sociology. Each science would, reflecting the model science of physics, be divided into a statics and a dynamics. Not all sciences had yet achieved their place in this classification system: sociology was still too inchoate, chemical dynamics were not yet clearly worked out. However, at that future time when all sciences did occupy their fated slots, there would no longer be any need for or possibility of a history of science. No need for because everything would flow forward from the first principles whose existence was guaranteed by the existence of the classification scheme. No possibility of because the stories to be told would have become so long and messy that students would get lost in the byways before they ever learned the first principles. Thus for a science-in-the-making we need a history, while a true science just needs to be part of a classification system. Comte still had a role for the great figures from the history of science: he developed a positivist calendar that replaced the old pagan names of the months and Christian saint's days with the names of great scientists. When geology no longer had a history, it would have a day and a month. So classification can often involve destruction of memory. As we have seen, Matsuda (1996, 109) interprets Clarapède's famous eyewitness experiment in just this way—what was remembered was not the person storming into the lecture hall but the type of person).

Cyberneticians have frequently announced the dawning of a new age and with it new classificatory principles. Thus, as we have seen, Pierre Auger declaimed that we have begun to live in the age of form. The old age, he argued, was one of diachrony and materialism: it gave us the historicist visions of Darwin and Marx (see Tort 1989 for diachronic classification in the nineteenth century). This age, he argued, is that of synchrony and form. When such an epistemic break is operated, the knowledge of the previous age becomes irrelevant; when the break is constituted by the move from diachrony to synchrony, the past is doubly deleted.

In Ross Ashby's work, among others, this new age then resonates with a need internal to cybernetic analysis to destroy memory. Consider the following extract from an article on general systems science as a new discipline, in which the need to do without memory is expressed purely as an internal concern:

> I have just said that when the Box is not completely observable, the Investigator may restore predictability *by taking account of what happened earlier.* Now this process of appealing to earlier events is also well known under another name. Suppose, for instance, that I am at a friend's house and, as a car goes past outside, his dog runs to a corner of the

room and cringes. It seems to me that the behavior is causeless and inexplicable. Then my friend says "He was run over by a car a month ago." The behavior is now accounted for *by my taking account of what happened earlier.*

The psychologist would say I was appealing to the concept of 'memory,' as shown by the dog. What we can now see is that the concept of 'memory' arises most naturally in the Investigator's mind when not all of the system is accessible to observation, so that he must use information of what happened earlier to take the place of what he cannot observe now. 'Memory,' from this point of view, is not an objective and intrinsic property of a system but a reflection of the Investigator's limited powers of observation. Recognition of this fact may help us to remove some of the paradoxes that have tended to collect around the subject. (Ashby 1958, 3)

In his *Introduction to Cybernetics*, Ashby railed against memory more technically but with an equally strongly weighted vocabulary, arguing that it was an epiphenomenal process that only our incomplete knowledge of the appropriate Markov chain for the given closed system (black box plus investigator) being considered. Memory is a metaphor needed by a ' "handicapped" observer who cannot see a complete system, and "the appeal to memory is a substitute for his inability to observe" (Ashby 1956, 115).

This theme of the destruction of memory is a complex one. It is not that past knowledge is not needed; indeed, it most certainly is in order to make sense of current actions. However, a *conscious* holding of the past in mind was not needed: the actant under consideration—a dog, a person, a computer—had been made sufficiently different that, first, past knowledge was by definition retained and sorted and, second, only useful past knowledge survived. Thus von Foerster's quantum theory of memory featured centrally an algorithm for forgetting (Beer 1959, 35–39). This position clearly resonates with Loewenberg's analysis of evolution we have looked at: you don't need a memory if evolution is doing your thinking for you. . . . And of course the way that evolution is doing this thinking is by forgetting: by deleting organisms that do not work. The creatures of evolution are hypotheses (Fogel, Owens, and Walsh 1965, 98).

One thing that becomes impossible when you deny the conscious holding of memory is any principle of duration: this ties back into the synchronic nature of cybernetic insights, the ability to shift between levels (one second can be a thousand years and so forth) and the denial of a difference between human and nonhuman actants. Past duration held within human memory is replaced by present emergence (Bergson's creative evolution) held in the configuration of objects and people in the now. Thus Bateson (1967) argued that "we are left regarding each step in a communicational sequence as a *transform* of the previous step. . . . We deal with event sequences which do not necessarily imply a passing on of the same energy" (32).

Retention of the past in consciousness would militate directly against the principle of feedback being applicable across the great divide between organic and inorganic. Thus Ross Ashby (1956), echoing Wiener, Bigelow, and Rosenblueth, argued that once it was appreciated that feedback could be used to correct any deviation: "it is easy to understand that there is no limit to the complexity of goal-seeking behaviour which may occur in machines quite devoid of any 'vital' factor. . . . It will be seen, therefore, that a system with feedback may be both wholly automatic and yet actively and complexly goal-seeking" (55). The future and teleology can be made respectively determinate and automatic, yet at the price of legislating away the past and history: by denying memory.

For Ross Ashby (1956), there was a clear methodological need to destroy one's memory of the past when applying cybernetic tools:

> Ordinarily, when an experimenter examines a machine he makes full use of knowledge 'borrowed' from past experience. If he sees two cogs enmeshed he knows that their two rotations will not be independent, even though he does not see them actually rotate. This knowledge comes from previous experiences in which the mutual relations of similar pairs have been tested and observed directly. Such borrowed knowledge is, of course, extremely useful, and every skilled experimenter brings a great store of it to every experiment. Nevertheless it must be excluded from any fundamental method, if only because it is not wholly reliable: the unexpected sometimes happens; and the only way to be certain of the relation between parts in a new machine is to test the relation directly. (19)

There is a triple destruction of memory implicit in this text. First, past disciplines are destroyed: they need to be created anew from first principles. Second, an individual experimenter must destroy his or her knowledge of previous experiments. Third, one result of this double destruction will be the discovery by cybernetics that memory itself is epiphenomenal. The three levels thus resonate, holding the same temporal pattern in different media. Note that this destruction of memory does not equate with the destruction of time and the observer in classical physics. In cybernetics, memory is destroyed so that history can be unified; in classical physics nonreversible time is destroyed so that history can be ignored.

Conclusion

It is not obvious that a social history of a universal discipline like cybernetics requires a reading of temporal patterns used in some of its texts—a fortiori in that I have not operated any kind of supervenient intellectual hierarchy to distinguish between good and bad cybernetics. I have argued that complex temporal patterning plays a central role in the elaboration of cybernetic texts in the period 1943–1975—and that, in particular, the destruction of memory is the temporal extension of the central notion of feedback. In order to sample this patterning, it was necessary to be able to range freely throughout the universe of cybernetic texts. The implicit social history comes out in the argument that the dissolution of boundaries between inside and outside that marked the formal operation of general systems analysis as developed by cyberneticians is complemented by a principle of resonance that makes all histories the same history. Thus, cybernetics operated in our period like a Möbius strip—if you follow round any narrative from the "inside" (say the working of the brain) you will get round to the "outside" (say, the history of humanity)—and of course vice versa. Like many universal disciplines through the ages, cybernetics offered a way of reading the macrocosm and the microcosm as reflections one of the other. Though the logic is the same, there was a difference in tendency from early forms that as a rule read humanity and human needs writ large in the external world: cybernetics tended rather to read the external world writ small in humanity.

Elizabeth Eisenstein (1979) points out that the printing press gave us a certain kind of packing algorithm for knowledge: the linear time of the narrative in coordinate space (left to right, top to bottom, forward in time as you read). In other words, one damn thing after another. In cybernetics, as in other universal disciplines, this kind of algorithm does not work: you need a principle for enfolding knowledge into itself. This enfolding is itself a very powerful tool; for a discussion of some of the roots of this power, see Lucien Dallenbach's analysis (1989) of the mirror in the text. One cannot understand cybernetics as a universal discipline without looking at the religious fervor and hypernatural excitement that it generates. One way of understanding these latter effects is in terms of the resonances that are set up. There is no such thing as a single cybernetic statement, true for only the level that it applies to and eschewing the three m's—metaphor, metonymy, and metaphysics. On the contrary, each statement in the universal discipline resonates at each possible level of analysis: the experiment, the experimenter, the history of the cybernetics, the history of the world. In this way a cybernetic text can be read as a verbal yantra: the complex enfolding of many registers of knowledge into a single textual space. In order for cybernetic interdisciplinarity to be achieved, all knowledge had to be in principle folded into the universal discipline. By looking at the temporal patterning of some cybernetic texts, we have seen how such enfolding is both possible and powerful.

Life, it has been said, is a message that the universe sends to itself. This statement encapsulates the kinds of messages that cyberneticians in the period

1943–1975 dealt in: messages that were exchanged and had meaning without for all that having conscious recipients processing them. Cybernetics was not so much a producer of Aesop's fables, where the moral is given at the end, but rather a producer of Sufi stories—which have no single moral, but you are changed by reading them. Wiener was right to reject "angels." He could—as Serres has done in a series of books—have chosen Hermes: a messenger who paradoxically is cloaked in secrecy and whose messages are about the way things are as much as they are about the state of mind of the gods (Serres 1980). He did well to choose "cybernetics," since the reference to feedback control implied messages with a distributed sender and without a past; with myriad recipients and marching sideways into the future.

Charles Lyell in the 1830s gave a sequential and jussive start to his archive. From this moment on (the publication of the *Principles of Geology*) all geological accounts should conform to the uniformitarian storytelling principles; in the past (before the archive), there was catastrophic succession, both in the earth through the stories told about it and in the haphazard nature of the progress of geology as a science. Now the twin temporalities would be synchronized, allowing the interpretive archivist a wide playing field in a progressive field. For the cyberneticians treated here, there is a similar sequential and jussive component to their memorial practices. Where you might have thought that you had a scad of new and distinct sciences (e.g., quantum mechanics and linguistics), you now—from this moment on—have to realize that you are dealing with one science (phase change in the atom is *about* communication (MacKay 1950)). Where you might have thought you were in the age of Empire, the age of physics, or the age of biology, you would as a cybernetician learn at the moment of your conversion that you always already were in the age of cybernetics. And stories that were to be stored in the archives—the stories that made *real* sense—would be stories of sameness: difference was to be banished as noise and distraction (with homeostasis providing the motor of history and classification its supercession). (Consider Bateson's ethnos or his schizophrenogenesis as ways of ensuring—as the Australian aborigines do in their dreamtime or as Joseph and his brothers do under Mann's treatement—that the same history can be told again and again and again, with difference just being in the detail of the instantiation of the mythical figure.)

Chapter 3 explores the interrelation of these two sets of memory practice in the memorializing of our ever-diminishing pool of available diversity. Before launching into—and roaming adrift on—another sea of data, a few remarks are in order. Are we in Lyell and the cyberneticians uncovering two analytically separate, historically unrelated memory practices each appropriate to the kind of scientific enterprise being considered and neither bearing much relationship to much anything else? Indeed, why are these things usefully called memory practices? They are both about ways of telling long-term histories. When we repeat to future generations the stories of the earth, of the cosmos, what do we need to remember about what went on before? These stories—in order to encompass vast time—attach to two distinct temporal and spatial topologies. Each fundamentally drew on an underlying discourse in some new infrastructural information technology for its time. Lyell offers a statistical, average history (so much land above the water on average all the time) that reads just as nicely as a meditation on the nature of record keeping in the early British Empire as it does an exposition of previous eons. The historical, geological truths he discovers are all the truths of the record-keeping process at that time being developed with enormous though rarely thoughtful energy all around him. Similarly, for the cyberneticians, the key to eternal, promiscuous recapitulation was the programmable state—literally in the case of Chile or the Soviet Union, or more generally as the black box representing any process, with sets of binary responses to inputs. The way they tell their histories is, as for Lyell, very readable as a meditation on the nature of a powerful new information technology: their histories became virtual machines that pushed the computer out into realms of the distant past and the insanely large, celebrating and exploring its power. And in the process, they were to colonize other disciplines—just as the computer did, by stating that what you thought had been a problem associated with your own specific tools and techniques is in fact a problem easily treatable by a cybernetician/computer. The two practices also both have strong mythical dimensions. Lyell was laboring against the seven days of Creation, the word of God itself about His own temporality. He offered the one way in which histories of the earth will henceforth be told, wrapped together with the one way in which histories of geology will be told, starting with Lyell. The cyberneticians (and Ashby's *Design for a Brain* is a great model for this) propagated a history of return together within an injunction to be in the present—to remember the process and not the specifics—and for many of them cybernetics became a charged, mythical calling in a way that few disciplines succeed. From the drear Golem in Wiener to the exuberant Beer in *Platform for Change* (1975), the raw power of their tools amazed them.

So we have three things in apposition: ways of writing histories of disciplines and of the earth/cosmos; meditations on infrastructural information technology; and mythically charged discourse. It is not surprising to find these three together in a set of memory practices: since the stories that we tell each other about our past constitute a thread that travels unbroken into a very distant past and is fundamental to all societies. Willy-nilly, an attempt to play with those stories is mythically charged. As a matter of historical observation (for what would an exact proof look like?), new memory practices in science and elsewhere accompany new information infrastructure at the same time as they engage in this mythic activity. The interesting question is perhaps what information infrastructure—how we gather and keep information about the world around us, in this case—has to do with our emergent globalizing ethnos: what is specific to our sets of stories as reflective of our information infrastructure and encapsulating our understanding of Old Time that makes them loosely unified, symbolically powerful, and policitically useful? And, as Latour reminds us, nevetheless and despite all avow that little excess known as universal truth: the *objet petit a* to our season of political jouissance as heirs to the earth. That is the central question of the next chapter.

3

Databasing the World: Biodiversity and the 2000s

Vaisampayana said:

The splendid monkey began to laugh and said, "Neither you nor anyone can see that form, for that was in another age that is no more. Time is different in the Eon of the Winning Throw, different in the Trey and the Deuce: this is the age of deterioration, and I no longer possess that form. Earth, rivers, trees, and mountains, Siddhas, Gods, and great seers adjust to time from eon to eon, as do the creatures: for strength, size and capacity decrease and rise again. Therefore enough of your seeing that form, scion of Kuru's line! I too conform to the Eon, for Time is inescapable."

Bhima said:

Tell me the number of Eons and the manner of each of them, the state therein of Law, Profit, and Pleasure, of size, power, existence, and death."

—Johannes Adrianus Bernadus van Buitenen, trans. and ed., *The Mahabharata*

Each plant fossil genus and species is to conform with the agreed format of 43 Fields for each Record defined by Holmes and Hemlsey (1990). Graphics may be included as part of the Record. The data will be stored on main frame and mini-computers as well as on personal computer discs. . . . A variety of database management packages are being tested. A specially customised version of *Smartware II* will be available commercially.

—Michael Charles Boulter et al., "The IOP Plant Fossil Record: Are Fossil Plants a Special Case?"

In the same way as sexual perverts, hysterics, or depressive maniacs, living creatures interacting with humans transform themselves to adapt to the new system represented by the labels. The real difference may be that life outside of human society transforms

Epigraphs: Johannes Adrianus Bernadus van Buitenen, trans. and ed., *The Mahabharata* (Chicago: University of Chicago Press, 1973), p. 504; Michael Charles Boulter et al., "The IOP Plant Fossil Record: Are Fossil Plants a Special Case?," in *Improving the Stability of Names: Needs and Options: Proceedings of an International Symposium, Kew, 20–23 February 1991,* edited by D. L. Hawksworth, (Königstein/Taunus, Germany: Koeltz Scientific, 1991), p. 240; Mary Douglas, *How Institutions Think* (Syracuse: Syracuse University Press, 1986), p. 101.

itself away from the labels in self-defense, while that within human society transforms itself towards them in hope of relief or expecting advantage."
—Mary Douglas, *How Institutions Think*

Introduction

The first two epigraphs—thresholds into a text, thresholds opening respectively into the annals of human mythology and into the archives of human knowledge—express radically different ways of thinking about the past. The past is a different country—they do things differently there (Hartley 1953): this is the clear and consistent message of *The Mahabarata.* The qualities that are cataloged to mark an Eon braid nature (for adherence to the laws of nature is part of the Law), sex, economy, and politics and precipitate all onto root ontological, eschatological questions of existence and death. Not a bad set to generate stories of radical singularity and mythical resonance with; and also not a bad set to explore nature with. Not our own though (where we is the globalizing "we" whose ethnos I will explore in the next three chapters). The Frankfurt declaration boldly asserts that all will be made the same—past eons will be colonized and incorporated using the tools of the bureaucratic present, just like any other darned other that confronts Western civilization. There is enormous power in being able to read the past, to tell the generations (Bowker 1994), and we are seeking to colonize the past just as we (the globalizing "we") are seeking to colonize the present. The third epigraph, from Mary Douglas, reminds us that there is a politics and a reality to the colonization. Entities become more and less real as they occupy different niches in the name game. Douglas is surely wrong about in her asymmetry: there are many humans who work to eschew labels (those labeled in a frenzy of moralizing wicked, terrorists, the Axis of Evil, perverted—the assorted demons of twenty-first-century capitalism), just as there are many non-humans who gravitate toward their label (Kohler's drosophili, mice strains, agricultural products). Her point is the stronger for its generalization. We conjure the world into a form that makes it (and us) manageable; this involves training entities into classifications.

Perhaps the most powerful technology—if only in English we could remove from the word *technology* its purely material dimension (Foucault's untranslatable "dispositifs techniques" are better)—in our control of the world and each other over the past two hundred years has been the development of the database. Starting with the rise of statistics (Porter 1986; Hacking 1990; Desrosières 1993) and the role of the archives commissioners in the 1830s through to the present, we find the ability to order information about entities into lists using classifications to be a contemporary key to both state and scientific power.

Databases are not a product of the computer revolution; if anything the computer revolution is a product of the drive to database (Campbell-Kelly 1994). The steps that were first taken in establishing the sway of precise knowledge about the past over actions in the present were more of the order of the development of manila folders and carbon paper in the nineteenth century (Yates 1989) than tied to the superb material act of inscribing slow memory into pulses of mercury as was the Williams tube in the 1950s. Over the past two hundred years, memory has gotten ever cheaper: a stick I keep in my pocket now contains more data than would a myriad Williams tubes or an office piled high with manila folders. Memory prostheses have multiplied around us; memory devices are legion; memory practices have been organized on massive scales: we now hold in the global memory details of events hundreds of thousands of years ago; masses of detail about lives in the twentieth century; countless photographs, tapes, diaries, souvenirs, memorials. . . . And yet paradoxically, many recent commentators have been more drawn to memory-politics (Hacking 1995) or archive fever (Derrida 1996); to social memory (Fentress and Wickham 1992) or invented tradition (Hobsbawm and Ranger 1992); to the memory of guilty actions (the Holocaust, slavery) or the largely unmediated (a cake does not really cut it here) remembrance of Proust's past (1989). The miracle of memory in our time is that memory practices are materially rampant, invasive, implicated in the core of our being and of our understanding of the world—and yet we experience them and discourse about them in terms of their ideal ramifications on some hypostasized entity created to void materiality from the equation: the individual, the nation-state, the people, and so forth.

And yet our memory practices are integrally material and symbolic.[1] Lyell and the cyberneticians have given us in turn two very material modes of memory practice: acting as archives commissioners or conjuring the world into a form that can be represented in a universal Turing machine whose past has been evacuated in order to render its future completely controllable. Integrally associated with each are two symbolic realms: memorializing difference and secular time through classification and hermeneutics, or memorializing sameness and circular time through abstraction and analysis.

We now move on to a consideration of the archival practices of scientists: how have they gone about trying to code and store the information that is necessary in order to replicate in small the archive of the history of the life. It is perhaps surprising that the appelation "natural history" has fallen into desuetude and indeed disfavor, in a world in which the task of producing theories

1. Thanks to Mike Cole here.

about historical time and its complexity is so crucial. What we need to understand about biodiversity is not how to freeze the present (the unfortunate connotation of "preservation" and "conservation"): indeed, it would be a disaster if we intervened to preserve all species at the point at which they had developed in the late twentieth century. We need to understand processes of change—rhythms of time and tangles of trajectories—in order to favorably inflect our futures. Just as geology was a leading science in the 1830s and computing was in the 1960s, so is biodiversity science—broadly conceived—today. We will uncover now a range of memory practices that tease apart the threads of memory as remembering what happened and memory as making present again. We seek to bank seeds, to database genomes, to clone our pets in a tumult of preservation of the Edenic present against the ravages of our sinful natures and Old Time. In so doing, we (the globalizing "we," that ethnos that does not have a tradition—our very knowledge is nontraditional) risk losing both past and future.

The Practice of Databasing the World

For the past few hundred years, many books and articles have begun with a phrase such as "We are entering a period of rapid change unimagined by our ancestors." The statement is both as true and as false now as it has been over the previous two centuries. It is true because we are as a society adjusting to a whole new communications medium (the Internet) and to new ways of storing, manipulating, and presenting information. We are, as Manuel Castells and others remind us (Castells 1996; Mansell et al. 1998), now in many ways an information economy, with many people tied to computers one way or another during our working day and in our leisure hours. It is false because we are faced with the same old problems—getting food, shelter and water to our human population; living in some kind of equilibrium with nature—as we ever were. How is the new knowledge economy impacting, and how can it potentially impact science and technology policy concerned with sustainable life?

Building an Infrastructure

Both standardization and classification are essential to the development of working infrastructures. Work done on standards committees and in setting up classification schemes is frequently overlooked in social and political analyses of technoscientific infrastructure, yet it is of crucial importance.

There is no question that in the development of large scale information infrastructures, we need standards. In a sense this is a trivial observation—

strings of bits traveling along wires are meaningless unless there is a shared set of "handshakes" among the various media they pass through. An email message is typically broken up into regular size chunks, then wrapped in various envelopes, each of which represent a different layer of the infrastructure. A given message might be encoded using the MIME protocol, which will allow various other mail programs to read it. It might then be chunked into smaller parts, each of which is wrapped in an envelope designating its ultimate address and its order in the message. This envelope is further wrapped in an envelope telling it how to enter the Internet. It might then quite possibly be wrapped in further envelopes telling it how to change from a configuration of electrons on a wire to a radio message beamed to a satellite and back down again. Each envelope is progressively opened at the end, and the original message reassembled from its contingent parts then appears "transparently" on your desktop. It's the standards that let this happen.

One observation that we can make at once is that it is standards all the way down: each layer of infrastructure requires its own set of standards. We might also say that it is standards all the way up. There is no simple break point at which one can say that communications protocols stop and technical standards start. As a thought experiment, let us take the example of a scientific paper. I write a paper about paleoecology for the journal *Science.* I know that my immediate audience will not be leading edge paleoecologists, but a wider community that knows only a little about Jurassic plant communities. So I wrap the kernel in an introduction and conclusion that discuss in general terms the nature and significance of my findings. A journalist might well write a more popular piece for the "Perspectives" section that will point to my paper. If this is my first paper for *Science,* then I will probably go through quite a complex set of negotiations through the peer review process in order to ensure the correct "handshake" between my message and its audience: Is it written at the right level? Have I used the right tone of authority? and so forth. Similar sets of standards/protocols arose with the development of office memoranda in the nineteenth century; they can also be seen with the development of email "genres" (Yates and Orlikowski 1992). I then need to be sure that the right set of economic agreements has been put into place so that my paper can be found on the Web by anyone at any time—where "at any one time" means the smooth time of our emergent ethnos, assuming paid access to the Internet, a relatively fast connection, and a recent version of Microsoft Word.

This common vision of disciplinary, economic, and network protocols serves the purpose of highlighting a central fact about infrastructures (Star and Ruhleder 1996). It is not just the bits and bytes that get hustled into standard form in order for the technical infrastructure to work. People's discursive and work practices get hustled into standard form as well. Working infrastructures standardize both people and machines (Woolgar 1987; Berg 1997). A further example will clarify this point. In order for the large-scale states of the nineteenth century to operate efficiently and effectively, the new science of statistics (of the same etymological root as the word *state*) was developed. People were sorted into categories, and a series of information technologies was put into place to provide an infrastructure to government work (regular ten-year censuses; special tables and printers; and, by the end of the nineteenth century, punch-card machines for faster processing of results). These standardized categories (male or female, professional, nationality, etc.) thus spawned their own set of technical standards (80 column sheets—later transferred to 80 column punch cards and computer screens . . .). They also spawned their own set of standardized people. As Alain Desrosières and Laurent Thévenot (1988) note, different categories for professional works in the French, German, and British censuses led to the creation of very different social structures and government programs around them. Early in the nineteenth century, the differences between professionals in one country or the other did not make so much difference. By the end of the century, these differences had become entrenched and reified; people became more and more like their categories.

At both the technical and the social level, there is no guarantee that the best set of standards will win. The process of the creation of standards for infrastructures is a long, tortuous, contingent one. The best-known stories here are the adoption of the QWERTY keyboard (good for its time in preventing keys jamming in manual typewriters; counterproductive now for most in that it places the bulk of the work onto the left hand—a hardship for many, but one appreciated by the southpaws (left-handers) among us—but so entrenched that there is no end in sight for it); the victory of the VHS standard over the technically superior Betamax standard; the victory of the DOS computing system and its successors over superior operating systems.

So why does the best standard not always win? There are two sets of reasons for this. First is that in an infrastructure you are never alone—no node is an island. Suppose there are five hundred users of DOS to every one user of a Macintosh. If I am a software developer, why should I write software for Macintosh computers, when I have a much smaller potential user base? So the strong get stronger. Going down one level, if I have to choose between writing an API (application program interface, to allow communication between one program and another) for interoperation between my program and an industry standard one or between my program and a rarely used one, it is clear which I will choose. More generally put, a new kind of economic logic has developed around the network infrastructures that have been created over the past two hundred years—the logic of "network externalities." The logic runs as follows. If I buy a telephone for $50 and there are only five people on the network, then it's not worth very much to me, unless I really like those five people. If five thousand more people buy telephones, then I haven't had to outlay another cent, but my phone is suddenly much more valuable. This situation describes *positive externalities.* De facto standards (such as DOS, QWERTY, etc.) gain and hold onto their position largely through the development of positive externalities. I buy a PC rather than a Mac because I know I will have access to the latest and best software. The second reason for the success of possibly inferior standards is that standards setting is a key site of political work. Arguably some of the most important decisions of the past fifty years have been made by standards-setting bodies: although one does not find a "standards" section of the bookshop alongside "history" and "new age" sections. Consider, for example, the "open source" movement. This movement has a long history, running deep into the origins of the Internet, proclaiming the democratic and liberatory value of freely sharing software code. The Internet, indeed, was cobbled together out of a set of freely distributed software standards. The open source movement has been seen as running counter to the dominance of large centralized industries—the argument goes that it puts power over the media back into the hands of the people in a way that might truly transform capitalist society. This promise of cyberdemocracy is integrally social, political, and technical. While there is much talk of an "information revolution," there is not enough discussion of the modalities through which people and communities are constituted by the new infrastructure.

Many models exist for information infrastructures. The Internet itself can be cut up conceptually a number of different ways. There is over time and between models a distribution of properties among hardware, software, and people. Thus one can get two computers "talking" to each other by running a dedicated line between them or by preempting a given physical circuit (hardware solutions) or by creating a "virtual circuit" (software solution) that runs over multiple different physical circuits. Or finally you can still (and this is the fastest way of getting terabits of data between two cities) put a hard disk in a truck and drive it down. . . . Each kind of circuit is made up of a different stable configuration of wires, bits, and people; but they are all (as far as the infrastructure itself is concerned) interchangeable (Tanenbaum 2003).

We can think of standards in the infrastructure as the tools for stabilizing these configurations. There is a continuum of strategies for standards setting. At the one end of the spectrum is the strategy of one standard fits all. This can be imposed by government fiat (e.g., the U.S. Navy's failed attempt to impose ADA as the sole programming language for its applications) or this can take the form of an emergent monopoly (e.g., Microsoft Windows XP/Vista). At the other end of the spectrum is the "let a thousand standards bloom" model. Here we enter the world of APIs and such standards as the ANSI/NISO Z39.50 information retrieval protocol. Z39.50 is a standard that has been developed for being able to make a single enquiry over multiple databases: it has been very widely adopted in the library world. We can use whatever database program we wish to make a bibliographic database, provided that the program itself subscribes to the standard. This means in essence that certain key fields like "author," "title," and "keyword" will be defined in the database. Now, instead of having to load up multiple different database programs in order to search over many databases (a challenge of significance well beyond the library community as large-scale heterogeneous datasets are coming into center stage in a number of scientific and cultural fields), we can frame a single query using the Z39.50 standard and have our results returned seamlessly from many different sources. The dream of XML is that it will create flexible and useful markup for texts and records. One sometimes hears about these two extremes that the former is the "colonial" model of infrastructure development while the latter is the "democratic" model. There is some truth to the implied claim that one's political philosophy will determine one's choice; and there is some solace for democrats in the observation that with the development of the Internet the latter has almost invariably won out—most Internet standards have been worked together in such a way that they permit maximal flexibility and heterogeneity. Thus, for example, if one looks at the emergence and deployment of collaborative computing, one finds that programs that try to do it all in one package (such as Notes) have been much less successful than less ambitious programs that integrate well with people's established computing environment: interoperability is the key.

Infrastructural development and maintenance requires work, a relatively stable technology, and communication. The work side is frequently overlooked. Consider the claim in the 1920s that with the advent of microfiche, the end of the book was nigh—everyone would have their own personal libraries (Abbott 1988); we would no longer need to waste vast amounts of natural resources on producing paper; all the largest library would need would be a few rooms and a set of microfiche readers. This is a possible vision—and one that we should not discount just because it did not happen (anyone who has used a microfiche reader will attest that it's a most uncomfortable experience—whereas if the same resources had gone into the failure as into the successful technology, it probably could have been a pleasure to work (see MacKenzie 1990 for a generalization of this theme of resource allocation). However, the microfiche dream, like the universal digital library, runs up against the problem that someone has to sit there and do the necessary photography/scanning; and this takes a huge amount of time and resources. It is easy enough to develop a potentially revolutionary technology; it is extremely hard to implement it—and even harder to maintain it.

Further, one needs a relatively stable technology. If one thinks of some of the great infrastructural technologies of the past (gas, electric, sewage, and so forth), one can see that once the infrastructure is put into place it tends to have a long life. Electrical wiring from before World War II is still in use in many households; in major cities, there is frequently no good map of sewage pipes, since their origin goes too far back in time. The Internet is only virtually stable—through the mediation of a set of relatively stable protocols (for this reason, it can be called an internetwork technology, rather than a network technology (Edwards 1998)). However, there is nothing to guarantee the stability of vast data sets. At the turn of the twentieth century, Paul Otlet developed a scheme for a universal library that would work by providing automatic electromechanical access to extremely well catalogued microfiches, which could link to each other (Rayward 1975). All the world's knowledge would be put onto these fiches, making his vision a precursor to today's hypertext. He made huge strides in developing this system (though he only had the person power to achieve a miniscule fraction of his goal). However, within forty years, with the development of computer memory, his whole enterprise was effectively doomed to languish as it does today in boxes in a basement—why retrieve information electromechanically using levers and gears when you can call it up at the speed of light from a computer? Much the same can be said of Vannevar Bush's never realized but inspirational vision of the Memex, an electromechanical precursor to the computer workstation. Large databases from the early days of the computer revolution are now completely lost. Who now reads punch cards—a technology whose first major use in the United States was to deal with the massive data sets of the 1890 census, and which dominated information storage and handling for some seventy years? Closer to home, the "electronic medical record" has been announced every few years since the 1960s—and yet it has not been globally attained (Gregory 2000). Changes in database architecture, storage capabilities of computers, and ingrained organizational practices have rendered it a chimera. The development of stable standards together with due attention being paid to backward compatibility provide an "in principle" fix to these problems. It can all unravel very easily, though. The bottom line is that no storage medium is permanent (CDs will not last anywhere near as long as books printed on acid free paper), so that our emergent information infrastructure will require a continued maintenance effort to keep data accessible and usable as it passes from one storage medium to another and is analyzed by one generation of database technology to the next.

In order to really build an information infrastructure, you need to pay close attention to issues of communication. We can parse this partly as the problem of reliable metadata. Metadata ("data about data") is the technical term for all the information that a single piece of data out there on the Internet can carry with it in order to provide sufficient context for another user to be able to first locate it and then use it. The most widespread metadata standards are the Dublin Core, developed for library applications, and the Federal Geographic Data Committee (FGDC) (http://fgdc.gov/) standard developed for Geographical Information Systems. If everyone can agree to standard names for certain kinds of data, then one can easily search for, say, authors over multiple databases.

Philosophically, however, metadata opens up into some far deeper problems. Take the example of biodiversity science. It is generally agreed that if we want to preserve a decent proportion of animal and floral life from the current great extinction event (one of six since the origin of life on this planet), then policymakers need the best possible information about current species size and distribution, as well as the ability to model the effect of potential and current environmental changes. In order to do this, you need to be able to bring together data from many different scientific disciplines using many different information technologies (from paper to supercomputer). But the point is that there is no such thing as pure data—as Hacking (1995) has shown, all categories come "under a description," and data comes in a dizzying set of categorical bins. You always have to know some context. And as you develop metadata standards, you always need to think about how much information you need to give in order to make your information maximally useful over time. And here we circle back to the first difficulty in developing an information infrastructure: the more information you provide in order to make the data useful to the widest community and over the longest time, the more work you have to do. Yet empirical studies have shown time and again that people will not see it as a good use of their time to preserve information about their data beyond what is necessary to guarantee its immediate usefulness—thus the medical culture of producing quick and easy diagnoses on death certificates in order to meet the administrative need and free time to get onto the next (live) patient (Fagot-Largeault 1989).

So standards are necessary—from social protocols down to wiring size. Indeed, when you foreground the infrastructure of your lives, you will see that you encounter thousands of standards in a single day. However, the development and maintenance of standards is a complex political and philosophical problem; and their implementation requires a vast amount of resources. Standards undergird our potential for action in the world, both political and scientific; they make the infrastructure possible.

Ownership of Scientific and Technological Ideas and Data

It has often been asserted that science is a public good: meaning that scientific work does not fit into the globally dominant market economy (Callon 1994). In the new knowledge economy, however, we are increasingly seeing the penetration of the market right down to the molecular level, right down to the stuff of scientific enquiry. Thus it is possible to patent genes, genetically modified plants, animals and so forth. Taking a fairly wide definition of ownership, we can see three main sets of issues arising from the implementation of this knowledge/information market: control of knowledge; privacy; and patterns of ownership.

By control of knowledge, I refer to the question of who has the right to speak in the name of the science. Since the mid-nineteenth century, this has been a fairly simple question to answer: only professionally trained scientists and doctors can speak for science and medicine in turn. Only they had access to the resources that were needed in order to speak authoritatively about a given subject; they had the journals, the libraries, the professional experience. Within the new information economy, this is not the case. For example, many patient groups now are being formed on the Internet. These groups often know more about a rare condition (e.g., renal cell carcinoma) than a local doctor does: they can share information twenty-four hours a day and can bring together patients from all over the world. This flattening out of knowledge hierarchies can be a very powerful social force. It carries along with it, though, the need to educate the enfranchised public about critical readership of the Web. There are many Web sites that look official and authoritative but in fact only push the hobbyhorse of a particular individual. We have through our schools and universities good training in reading and criticizing print media, but we have little expertise as a culture in dealing with highly distributed information sources.

Privacy concerns are a significant dimension of science and technology policy in the new economy. It is now technically possible to generate and search very large databases, and to use these to integrate data from a whole series of domains. As this happens, the potentialities for data abuse are increasing exponentially. Much has been written, for example, about data mining of the Icelandic population. After much public debate, citizens of Iceland agreed to sell medical and genealogical records of the country's 275,000 citizens to a private medical research company. There were two central reasons for choosing

Iceland: it has a population that has a relatively restricted gene pool; and it has excellent medical and parish records dating back some thousand years. While the science may prove useful (the question is open), it certainly raises the specter of genetic screening of prospective employees by a given company. It is extremely difficult to keep records private over the new information infrastructure—many third-party companies, for example, compile together data from a variety of different agencies in order to generate a new, marketable form of knowledge. There is no point in trying to adhere to the old canons of privacy; however, open public debate and education about the possibilities of the new infrastructure are essential.

Further, there are new patterns of ownership of information/knowledge. Science has frequently been analyzed as a "public good," according to which it is in the interests of the state to fund technoscientific research since there will be a payoff for society as a whole in terms of infrastructural development. With the increasing privatization of knowledge (as we turn into a knowledge-based economy), it is unclear to what extent the vaunted openness of the scientific community will last. Many refer to a "golden age" when universities were separate from industry in a way that they are not today. While a lot of this talk is highly exaggerated (science has always been an eminently practical pursuit), it remains the case that we are in the process of building new understandings of scientific knowledge.

A key question internationally has been that of who owns what knowledge. This is coming out in fields like biodiversity prospecting, where international agreements are in place to reimburse "locals" for bringing in biologically active plants and so forth. However, the ownership patterns of knowledge of this sort are very difficult to adjudicate in Western terms. For example, consider a Mexican herbalist selling a biologically active plant in a market in Tijuana. He owns the plant but is not the source of knowledge about biologically active plants. This knowledge does not go back to a single discoverer (as is needed in many Western courts of law adjudicating matters of ownership of intellectual property) but to a tradition held, often, by the women of a collectivity. The herbalist may well not be able to trace back the chain of ownership that goes back to the original harvesting of the specific he or she is selling (Hayden 1998). Similarly, Australian aborigines or the Native Americans had very different concepts of land ownership from the white settlers; leading to complex negotiations that continue today about the protection of natural resources (Watson-Verran and Turnbull 1995). We need anthropological/sociological studies of local knowledge (to the extent to which this is being mined by scientists) again in order to help design just frameworks and studies of issues of data ownership in different countries. There is a danger when we talk about the explosion of information in the new knowledge economy that we forget the role of traditional knowledge in the development of sustainable policies for a region. Thus research has shown that management of some parks in the Himalayas has relied on models brought in from the outside and taught to villagers through the distribution of television programs, while at the same time ignoring centuries of local ecological knowledge because it is practice-based, has its own intricate weaving of knowledge about the environment, religious belief, and mythological expression, and cannot be easily conjured into a form that can be held on a computer (Padel 1998).

Sharing Data

The form of scientific work most studied by sociologists of science is that which leads from the laboratory to the scientific paper by means of the creation of ever more abstract and manipulable forms of data, which Bruno Latour (1987) has dubbed "immutable mobiles." In this process, there is no need to hold onto data after it has been enshrined in a scientific paper: the paper forms the "archive" of scientific knowledge (frequently adopting names redolent of this storage ambition, such as the *Archives for Meteorology, Geophysics and Bioclimatology*). The scientific paper, which is the end result of science, contains an argument about a hypothesis (which is proved or disproved) and a set of supporting data that is, saving a controversy, taken on faith by the scientific community. The archive of scientific papers can then be indexed in terms of both arguments made and information stored.

However, over the past twenty years, we have seen in a number of new and of formerly canonical sciences a partial disarticulation of these two features of scientific work. Increasingly, the database itself (the information stored) is seen as an end in itself. The ideal database should according to most practitioners be theory-neutral, but should serve as a common basis for a number of scientific disciplines to progress. Thus one might cite the human genome initiative and other molecular biological projects as archetypical of a new kind of science in which the database is an end in itself. The human genome databank will in theory be used to construct arguments about the genetic causation of disease, about migration patterns of early humans, about the evolutionary history of our species; but the process of producing causation is distinct from the process of "mapping" the genome—the communities, techniques, and aims are separate.

This disarticulation, which operates in the context of producing a working archive of knowledge, is not in itself new. To limit ourselves arbitrarily to the past two hundred years, a significant percentage of scientific work has been involved with creating such an archive. Napoleon's trip to Egypt included a boatload of geologists, surveyors, and natural historians and reflected a close connection between the ends of empire and the collection of scientific knowledge (Nolin 1998). The same went for Smith's geological survey of Britain (Winchester 2001) and Cook's travels to Australia. Richard's *The Imperial Archive* (1996) presents some wonderful analysis of the imperial drive to archive information in order to exercise control (a theme familiar to readers of Latour). The working archive is a management tool. What is new and interesting is that the working archive is expanding in scale and scope. As French philosopher Michel Serres (1990) points out, we are now as a species taking on the role of managing the planet as a whole—its ecosystems and energy flows. We now see nature as essentially only possible through human mediation. We are building working archives from the submicroscopic level of genes up through the diversity of viral and bacterial species to large-scale floral and faunal communities and the mapping of atmospheric patterns and the health of the ozone layer. There is an articulation here between information and theory, but the stronger connection is between information and action—with competing models based on the same data-producing policy recommendations. In this new and expanded process of scientific archiving, data must be reusable by scientists. It is not possible to simply enshrine one's results in a paper; the scientist must lodge her data in a database that can be easily manipulated by other scientists.

In the relatively new science of biodiversity, this data collection drive is reaching its apogee. There are programs afoot to map all floral and faunal species on the face of the earth. In principle, each of these maps should contain economic information about how groups of animals or plants fend for themselves in the web of life (http://curator.org/WebOfLife/weboflife.htm) and genetic information (about how they reproduce). In order to truly understand biodiversity, the maps should extend not only out in space but back in time (so that we can predict how a given factor—like a 3 degree increase in world temperature—might effect species distribution). Large-scale databases are being developed for a diverse array of animal and plant groups and the SPECIES 2000 program of the International Union of Biological Sciences (IUBS), CO DATA, and the International Union of Microbiological Societies (IUMS) has proposed that they might eventually be merged into a single vast database of all the world's organisms—though there are competing large lists (http://www.all-species.org/). NASA's Mission to Earth program is trying to "'document the physical, chemical, and biological processes responsible for the evolution of Earth on all time scales.'" The U.K. Systematics Forum publication *The Web of Life* quotes E. O. Wilson's invocation: "'Now it is time to expand laterally to get on with the great Linnaean enterprise and finish mapping the biosphere,'" and speaks of the need to "discover and describe the Earth's species, to complete the framework of classification around which biology is organized, and to use information technology to make this knowledge available around the world'" (http://www.nhm.ac.uk/hosted_sites/uksf/web_of_life/index.htm). These panoptical dreams weave together work from the small-scale molecular biological to the large-scale geological, and temporally from the attempt to represent the present to a description of the history of all life on earth. They constitute a relatively direct continuation of the drive for the imperial archive. While the notional imperial archive sought to catalog completely the far-flung social and political empire in order to better govern it, biodiversity panopticons seek to catalog completely the natural empire for much the same reason. Although they cover only a thin slice of species and environments (and so constitute oligoptica; see Latour and Hermant 1998), they are created to be, and are manipulated as if they were, panopticons.

The information collection effort that is being mounted worldwide is indeed heroic. Databases from far-flung government agencies, scientific expeditions, and amateur collectors are being integrated more or less successfully into large-scale searchable databases. Science and technology policy analysts have a significant contribution to make to the process of federating databases in order to create tools for planetary management. We can produce means to engage the complexity and historicity of data within the sciences so that social, political, and organizational context is interwoven with statistics, classification systems, and observational results in a generative fashion. We need to historicize our data and its organization in order to create flexible databases that are as rich ontologically as the social and natural worlds they map, and that might really help us gain long-term purchase on questions of planetary management.

Even if we can name everything consistently, there are the problems of how to deal with old data and how to ensure that one's data doesn't rot away in some information silo for want of providing enough context. The problem with much environmental data is that the standard scientific model of doing a study doesn't work well enough. In the standard model, one collects data, publishes a paper or papers, and then gradually loses the original data set. A current locally generated database, for example, might stay on one's hard drive for a while then make it to a zip disk; then when zip technology is superseded, it will probably become for all intents and purposes unreadable until one changes jobs or retires and throws away the disk. There are a thousand variations of this story being repeated worldwide—more generally along the trajectory of notebooks to shelves to boxes to dumpsters.

When it could be argued that the role of scientific theory as produced in journals was precisely to order information—to act as a form of memory bank—this loss of the original data was not too much of a problem. The data was rolled into a theory that not only remembered all its own data (in the sense of accounting for it and rendering it freely reproducible) but potentially remembered data that had not yet been collected. By this reading, what theory did was produce readings of the world that were ultimately data independent; if one wanted to descend into data at any point, all one had to do was design an experiment to test the theory and the results would follow.

However, two things render this reading of the data/theory relationship untenable. First, it has been shown repeatedly in the science studies literature that scientific papers do not in general offer enough information to allow an experiment or procedure to be repeated. This entails that in a field where old results are continually being reworked, there is a need to preserve the original data in as good a form as possible. Second, in the biological sciences in general and the environmental sciences in particular, the distributed database is becoming a new model form of scientific publication in its own right. The Human Genome Initiative is resulting in the production of a very large collaborative database, for example. In the environmental sciences, where the unit of time for observing changes can be anything from the day to the millennium, there is a great value in having long, continuous data sets. The problem of what data to retain in order to keep a data set live is a metadata problem; and as Ingersoll et al. (1997) note in "A Model Information Management System for Ecological Research": "the quality of metadata is probably the single most important factor that determines the longevity of environmental data" (310).

Science is an eminently bureaucratic practice deeply concerned with record keeping, as Latour (1987) reminds us in *Science in Action*. Disciplines do a mixed job of keeping track of their own results over time; indeed, a key finding of science studies has been that using "theory" as a way of storing old, and accounting for potential, data can be highly problematic since replacement theories do not automatically account for all the data held in the outgoing one (the locus classicus is Kuhn's book on *The Structure of Scientific Revolutions*, first published in 1962 (Kuhn 1970)). The difficulties become apparent when you move beyond the arrangement and archiving of data within a given science to look at what happens in the efforts of a vast number of sciences (working from the scale of molecular biology on up to that of biogeography or even cosmology) to coordinate data between themselves within the field of biodiversity. In practice, the sciences use many differing "filing systems" and philosophies of archival practice. There is no automatic update from one field to a cognate one, such that the latest classification system or dating system from the one spreads to the other. Further, it is often a judgment call whether one needs to adopt the latest geological timeline, say, when storing information about ecological communities over time—particularly if one's paper or electronic database is structured in such a way that adopting the new system will be expensive and difficult. Such decisions, however, have continuing effects on the interpretation and use of the resultant data stores.

There has been relatively little work dealing with the organizational, political, and scientific layering of data structures. It is clear, though, that the assignation of an attribute to the world of discourse or of materiality is shifting, post hoc. Information infrastructures such as databases should be read both discursively and materially; they are a site of political and ethical as well as technical work; and that there can be no a priori attribution of a given question to the technical or the political realms.

This means practically that it is a policy priority to pay attention to the work of building large-scale databases, or of developing large-scale simulations. It is no longer the case that knowledge held in a particular discipline is enough to carry out scientific work. From the 1940s on (with the Manhattan project) one might say that large-scale technoscience is inherently massively multidisciplinary. However, scientists are not trained to share information across disciplinary divides. And computer scientists cannot do the work of translating between disciplines. Indeed, one of the major difficulties with developing new scientific infrastructures using computers is that the work that is interesting for the computer scientist is often very high-end: involving, say, the latest object-oriented programming and visualization techniques. However, the work that is important for the scientist might be theoretically uninteresting for the computer scientist: for example, producing good ways of updating obsolete databases. There are two sides to the solution here. One the one hand, career paths must be developed that are more in tune with the needs of technoscience. This has worked successfully with the training of a cadre of bioinformaticians with the human genome project at the University of Washington. This cadre knows both molecular biology and computer science—and has a possible career path outside of the confines of the traditional disciplinary structure. On the other hand, we need to put the maintenance of the information infrastructure high on the agenda. Many scientists will go for grants to get the latest equipment; few will concern themselves with upgrading old databases or reusing old data. Huge amounts of data are being lost this way—data about the effects of human activity on this planet that is essential if we are to build a workable future. Just as software reuse has become the clarion call of the latest revolution in programming techniques, so should data reuse become a clarion call within technoscience.

International Technoscience

There has been much hope expressed that in the developing world, the new information infrastructure will provide the potential for a narrowing of the knowledge gaps between countries. Thus an effective global digital library would allow Third World researchers access to the latest journals. Distributed computing environments (such as the Grid, being developed in the United States) would permit supercomputer grade access to computing to scientists throughout the world. The example of the use of cell-phone technology to provide a jump in technology in countries without landlines has opened the possibility of great leaps being made into the information future. As powerful as these visions are, they need to be tempered with some real concerns. The first is that an information infrastructure like the Internet functions as a Greek democracy of old: everyone who has access may be an equal citizen, but those without access are left further and further out of the picture. Second, access is never really equal: the fastest connections and computers (needed for running the latest software) tend to be concentrated in the First World. This point is frequently forgotten by those who hail the end of the "digital divide"; they forget that this divide is in itself a moving target. Third, governments in the developing world have indicated real doubts about the usefulness of opening their data resources out onto the Internet. Just as in the nineteenth century the laissez-faire economics of free trade was advocated by developed countries with the most to gain (because they had organizations in place ready to take advantage of emerging possibilities), so in our age the greatest advocates of the free and open exchange of information are developed countries with the most robust computing infrastructures. Some in developing countries see this as a second wave of colonialism—the first pillaged material resources, and the second will pillage information. All of these concerns can be met through the development of careful information policies. There is a continuing urgent need to develop such policies.

International electronic communication holds out the apparent promise of breaking down a first world/third world divide in science. With developments like the remote manipulation of scientific equipment such as the SPARC project, whose scientists on the Internet can manipulate devices in the Arctic Circle without having to go there. The possibility of attending international conferences virtually also exists. And if universities succeed in wresting control over scientific publications from the huge publishing houses (a very open question), then easy/cheap access to the latest scientific articles becomes possible for a researcher in outback Australia. At the same time, a number of forces are working to reinforce the traditional center/periphery divide in science internationally. Even with the move to open up access to scientific publications and equipment, there is no guarantee that the "invisible colleges" (Crane 1972)—which operate informally and determine who gets invited to which conference and so forth—will change: indeed, the evidence seems to be to the contrary. Further, at the current state of technological development, there is a significant gap between information access in different regions of any given country, or in different parts of the world. Consider the analogy of the telephone. In principle, anyone can phone anywhere in the world; in practice, some regions have more or less reliable phone services, which may or may not include access to digital resources over phone lines. In fact, half of the world's population does not have telephone access.

We can go beyond the continuing digital divide, however, to consider the possibility of mounting large-scale scientific data collection efforts. Such efforts are central to the social sciences, and to the sciences of ecology and biodiversity. With the development of handheld computing devices, it is becoming possible for a semi-skilled scientific worker with a minimum of training to go into the field and bring back significant results. Thus in Costa Rica, the ongoing attempt to catalog botanical species richness is being carried out largely by "parataxonomists" who are provided with enough skills in using interactive keys (which help in plant recognition) to carry out their work almost as effectively as a fully trained systematist (http://www.inbio.ac.cr/es/default.html). Computer-assisted workers, together with the deployment of remote sensing devices whose inputs can be treated automatically, hold out the possibility of scaling up the processes of scientific research so that they are truly global in scale and scope.

Distributed Collective Practice

Collaborative work is central to the new knowledge economy. Traditionally, scientific breakthroughs have been associated with particular laboratories—the Cavendish laboratory in Cambridge, England, for example. Such laboratories have always been a site for collaboration, in the form of exchanging ideas with visiting scholars, holding conferences, and training graduate students. However, it is impossible nowadays to imagine managing a large-scale scientific project without including far-flung collaborators—particularly if one is seeking to develop a truly global scientific culture.

Although technoscientific work is inherently collaborative, management structures in universities and industry still tend to support the heroic myth of the individual researcher. Many scientists turn away from collaborative, interdisciplinary work—precisely the kind of work that is most needed in order to develop policies for sustainable life—because they are risking their careers if they publish outside their own field. There is significant institutional inertia, whereby an old model of science is being applied to a brave new world.

This is coming out most clearly in the area of scientific publications. In the field of physics and then in a number of other scientific disciplines, we are witnessing the spread of electronic preprints and electronic journals. Traditional academic journals run by huge publishing conglomerates just cannot turn around papers quickly enough to meet the needs of scientists working in cutting-edge fields. Throughout the past two centuries, there has been a relatively stable configuration whereby journal articles have become the central medium for the dissemination and exchange of scientific ideas. In principle there is no reason why a scientist today should not publish her findings directly on the Web. As in many sectors of the new information economy, the development of the new publication medium is leading to a reconsideration of just what kind of value the large publishing houses add to journal production. As more and more journals go online, they are being forced to go beyond their traditional service of providing distribution networks and find ways to bring their material onto the web. It seems likely that all major scientific journals will soon be accessible on the web; even though the economics of such distribution is not yet fully worked out: possibilities include paying for each paper downloaded, or buying an institutional subscription for a whole journal. A more far-reaching implication is that the journal article may no longer function as the unit of currency within the research community.

The policy implications are clear. Great attention must be paid to the social and organizational setting of technoscientific work in order to take full advantage of the possibilities for faster research and publication cycles. There is a well known paradox about the development of computing—the productivity paradox—according to which the introduction of computers into a workplace tends to lead to a lowering of productivity in the short term (about 20 years). Paul David (1985) and others have argued that what is happening here is that we are still using the old ways of working, and trying to adapt them to the production of electronic text.

The choices that we are making now about the new information infrastructure are irreversible. The infrastructure is performative (in that it shapes the forms that technoscience will take,) and it is diffuse (there is no central control). There is currently widespread belief in technical fixes for inherently social, organizational, and philosophical problems—such as curing the ills of incompatible data sets through developing metadata standards. Furthermore, there is a disjunct between the policy and the informatics discourses. Major works on the politics of environmental discourse do not mention environmental informatics at any point: they write as if there is no layer at all between science and politics.

We are globally faced with problems that cannot be solved by the generation of knowledge alone. Producing massive lists of flora and fauna will not of itself lead to a way to preserve biodiversity. However, the information economy does promise some major new tools. The real key to developing technoscientific policy in the context of the new knowledge economy is operating a deep understanding of the nature of information infrastructures. For many, the development of science policy evokes ethical and political questions such as genetic screening, cloning and so forth. This is important work, but equally important is the work of monitoring the standards and classification systems that are getting layered into our models and simulations and that of changing our institutions so that they can take maximum advantage of the new collaborative and information-sharing possibilities.

The central issues for science and technology in the context of the new knowledge economy are the development of flexible, stable data standards; the generation of protocols (both social, in the form of international agreements about data exchange, and technical, in the sense of metadata standards) for data sharing; and the restructuring of scientific careers so that building large-scale scientific infrastructures is as attractive a route as performing high-profile theoretical work. The necessary tools are being created, but they are widely scattered and often lost. The new information infrastructure for technoscience will be extremely powerful; with good bricolage, we can make it just and effective as well.

These data standards and protocols will provide enormous material power to our emergent globalizing civilization; they will allow us to act in the present in ways we cannot yet imagine. And yet, as we will see, the work of producing these databases is inherently political and philosophical and about our relationship with our past despite their acclaimed practicality in the present. What we choose to remember about the past and our ways of remembering shape our appreciation of and actions in the present. We aim to create (we call it "preserve," "conserve," "maintain") in the future a world whose history will be the history recorded in our databases. Exclusion from databases has drastic consequences for entities, to say nothing of "systems and services" in current environment-speak; you can only protect through policy interventions that which can be named, that which can be shown to have been important in the past. What we are doing now—globally, willy-nilly—is setting the agenda for what the world will be based on our understand of what the world has been. Thus it is critical at this juncture that we pay close attention to the ontologies and politics of our databasing of life.

Accounting for Life

In general, there has been just as much irregularity in the archival practices of scientists as they in turn have noted in the earth. Thus Richard Pankhurst (1993), developer of the pioneering PANDORA taxonomic database, wrote:

The normal practice of taxonomists when preparing an account of the plants of a region (a flora) or animals of a region (a fauna) or a monograph of a particular genus or family has been to collect the relevant data over a long period, and at the end to publish only some of the most vital facts together with various conclusions or assertions. The main collection of data on which the research was based is then just shelved, and often lost or forgotten, or even deliberately destroyed. The time period between revisions of a flora or monograph may be as long as a hundred years, so that the next worker to take up the subject often has to reassemble the data from scratch. This is of course tremendously wasteful, and may be poor scientific practice, as well. (299)

This passage is formally analogous to Lyell's description of the archival practices of the earth itself: only a portion of the needed information is preserved; the record is irregularly revised, and there is no guarantee of comparability of results from one era or area to the next. Indeed, as leading botanical informatician F. A. Bisby (1988) remarked, it is unclear just where the archive is—and what form it takes:

The most clear-cut internal communications concern three support services that are used throughout the taxonomic profession: nomenclatural indexing, bibliographic records and collections of biological materials. A less distinct area is that of taxonomic revision; in the long-term do the monographic revisions determine the framework of classification or do the numerous changes, either by the authors of floras, faunas, and handbooks, or by local taxonomists describing new forms add up to an equally important contribution? A body of information is built up either of a single agreed system or for competing views in a group, but where does this information presently reside? Is it to be found in the heads of the relevant taxonomic experts or does it reside in the nomenclatural indexes, the literature, and the labeling of materials in collections? Finally, there are the 'products', floras, faunas, handbooks, checklists, and occasionally monographs, to go out into the public to provide, along with expertise and education, the external communications described above. (284)

The problem of the where and what of the archive is a common one across many sciences.

For the past twenty years, much has been made of the possibilities of using computing technology to make the archive more regular and accessible. Computers figure in our story in several ways. For our purposes here, we will concentrate on how database technology has changed the ways in which classifications can be developed. It is worth stepping back a while to look at the general history of classifications in order to make the point that I am not here giving the contingent history of a particular discipline or narrow set of disciplines, but rather I am describing a general change that has occurred in many fields with the introduction of the computer.

Going back to the grandfather of modern taxonomy—Linnaeus—we see the importance of the human memory as a storage device. In the eighteenth and nineteenth centuries, the natural historian or geologist would go to the field armed with classification systems and books. In a classic article on logic and memory in Linnaeus's system of taxonomy, A. J. Cain (1958) noted: "Linnaeus states explicitly and repeatedly that the botanist (and, since his rules apply equally to all kingdoms of Nature, the zoologist too) must know all genera, and commit their names to memory, but cannot be expected to remember all specific names" (156). This requirement had clear consequences for Linnaeus: the names of genera, in order to be mnemonically memorable, had to be univerbal, distinct, literally meaningful (not meaningless and therefore hard to retain in memory), expressive of the chief characteristic of the genus and not contradicted by the feature of any species in the genus (ibid.). Information down to the species level could not be kept in the head, and so the book was taken into the field: some of the earliest books ever printed were field guides for naturalists.

During the later part of the nineteenth century, as Chandler (1977) and then Yates (1989) have so beautifully described, new information technologies developed which permitted the easy storage and coding of larger amounts of data than could previously be easily manipulated. This was a response to a hugely increased demand for information processing across the board—in banking, insurance, public health, international labor, geology, and biology. They were used both in the business of government and in the business of science: just as statistical thinking spanned both fields in the 1830s. In the 1890s, as Patrick Tort (1989) has pointed out, there was a positive explosion in the number of international classifications that were developed: among the things classified were types of work, shapes of the ear (Bertillon), criminal physiognomy (Lombrosio), and disease (Bertillon). (Tort 1989) As these classifications came into play, they drew on currently available paper technology from whatever field—for example, there were two hundred disease categories in the original International Classification of Diseases since this was the number of rows in the standard Prussian census form (Bowker and Star 1999).

There resulted a kind of international stability in classification systems; however, the resultant archives were rigid. Consider the problem of the classification of plants. In a paper-based archival practice, different "type specimens" (instances of a particular plant that were taken to define the species) would be stored together on shelving according to the filiation that they had with other plants as specified by the version of the classification system then operating. If a plant was for whatever reason reclassified, the type specimen had to be physically moved from one location to another, a set of new index cards had to be typed, old index cards needed to be removed or relocated, and series of articles or books quite possibly had to be physically relocated as well. The order of the archive in theory matched the understanding of the order of nature. Clearly, over a whole range of classification systems, this led to enormous inertia; it was physically difficult to rearrange all the available specimens and pieces of paper and books, regardless of the validity of the suggested change. (This is akin to the museological paradox of whether to hold onto an old organization of, say, a natural history collection with its own historical interest or reordering the exhbits in line with more recent theory. With the development of computing, it was suggested that classifications could become more flexible. In the classification of viruses, for example (notoriously difficult since viruses are highly changeable over time and between hosts), it was suggested that a form of multifaceted classification system could be used, with a series of defining physical and chemical characteristics being tagged (Matthews 1983; Murphy et al. 1995). This would free the virologist from having to determine at any one moment that such and such a virus belonged in the same family as some other; family affiliation could change with the vagaries of theory without affecting the classification system, and thence the location of paper and books. This form of multifaceted classification was also developed in the context of computer applications for medicine (the Read classification).

Here again, the problem of being able to efficiently retrieve information about past diseases would in theory be made much easier to deal with if instead of statistically lumping together an unknown disease (e.g., AIDS in the 1970s) with pneumonia or cancer, the disease would be defined by a set of parameters, whose regularities could later be teased out and explained with the development of the theory of slow viruses (Grmek 1990). And, indeed, multifaceted classifications have been widely adopted within the field of library science as providing a degree of flexibility unimagined with single universal classification schemes. So the change has been from one theory of classifications embedded in one kind of paper-based record-keeping practice to a new theory embedded in a computer-based archival practice.

This has rendered the past in theory much more accessible—by allowing a reinterpretation on the fly of past archives. The development of new database technologies have brought this goal closer to hand. The first set of databases historically introduced were hierarchical (e.g., include IMS from IBM, available in 1968) (Khoshafian 1993). In the field of floristics, for example, the Flora of Veracruz, initiated in the mid-1960s, was designed to create a database of specimen labels but was then maintained to curate the herbarium at Jalapa. These technologies allowed a record type—in our examples, a plant—to have a single "parent": that is to say, the system modeled whichever hierarchical classification was in use in the field being represented. Computer scientist E. F. Codd in the early 1970s introduced the relational data model that engineered a split between the physical storage of data in the computer and the representation of that data (Khoshafian 1993, 20). Each record might be stored in the system by its record number—the order in which it was entered. However, at any point, the user of the system could specify a set of relationships that it was interested in (e.g., leaf size and color) and produce a "virtual" archive that reflected that set of relationships. As Pankhurst (1993) notes, this did not solve the problem of a hierarchical classification enforcing a particular reading of the past unless there was careful design of the database: "It is frequent practice to have one record for each species and to have fields within each record for other categories such as family, genus, or subspecific taxa. This is not actually wrong, but it does have an important disadvantage, which is that it becomes awkward to introduce other levels of taxonomic category as an afterthought. A better way to express this kind of data is instead to have one record for every name, regardless of its taxonomic level" (231). His PANDORA database was designed with this injunction in mind; as was the TROPICOS database used by the Missouri Botanical Gardens (responsible for the Flora of North America and the Flora of China). The most recent generation of databases, object-oriented ones, allow a greater degree of flexibility again by making it easy to redefine basic operations on the fly, so that not only the ordering of the archive but also its manipulation can be reconfigured according to present theory.

We are by no means in a position where the past can be reconfigured at will. It has been argued by some systematists concerned with the general problem of the fossil record that fossils should not enter into contemporary classifications; because they tend to inconveniently force the insertion of a new taxonomic level between two old ones, thus forcing a reworking of the complete system (Scott-Ram 1990). N. F. Hughes (1991) remarks:

> The use of fossils has in recent years come to be regarded as cumbersome and unproductive; the work is said to abound with unimaginative complacency, with the obscurity of esoteric terminology, and with lack of compatibility of treatment of different groups including even that between the fossils of plants and animals. As a result, much effort has been directed by geologists in solving their stratal problems towards employing any other available physical or chemical phenomena, and thus to avoiding altogether the 'expensive' and supposedly ineffective use of palaeontologists and their fossils. (39)

Hughes describes petroleum geologists developing their own heuristic classifications for their databases. Furthermore, and most important, we need to battle the stereotype common to computer textbooks of an advancing wave of history. There are a lot of "obsolete" databases using "old" biological and geological classification systems in different areas, and these old forms are doing good work in the present. With so much hyperbole attached to writing about computers, we tend to reify ghostware (the afterlife of vaporware—written about as if it already existed and was used, and becoming more real over time).

The manipulation of the archive itself has been strongly affected by the development of computing technology. To look at this issue, we will briefly consider the recent history of cladistics—one of the main schools of biological classification. There is no need here to go into full detail about the nature of cladistics—the reader is referred to the works of Scott-Ram (1990), Ridley (1986), and Hull (1966) and to the journal *Cladistics* (it is worth going back to the source here since there were very strong criticisms in that journal about all of these general references) for more information. For our purposes, it suffices to say that cladistics, which grew out of Willi Hennig's work, developed out of the older school of "evolutionary taxonomy." In the latter, the classification system was seen as producing a best guess about the evolutionary history of a particular organism. In so doing, it would draw on a series of techniques and produce a reasoned judgment over the set of them. Cladistics gave a more regular algorithm for determining phylogeny (the past of a given phylum), essentially by focusing attention on shared, derived characteristics of sets of organisms and by using "outgroup" comparisons (comparisons of sets of organisms sharing many features but not some particular derived characteristic) in order to develop the classification system. It stood in the early years in opposition to "numerical taxonomy," which did not attempt to represent the history of the organism in the classification system, but rather to seek to determine the optimal number of characters that could be used to distinguish one species from the next. Finally, there were the "pattern cladists" who deployed cladistic techniques without tying them to any interpretation of the past, although the very existence of such a school of thought was denied by some who were claimed by critics to be its chief representatives.[2] In a typically

2. "How then did the prevailing superstition about pattern cladists become established? The answer, I suggest, is to be found in the uses to which the myth has been put. Consider as an analogy Halstead's (1980) charge that cladistics is a communist conspiracy. On its face it seems simply a mindless lie, but in fact it has quite a different purpose than to represent an actual state of affairs. Halstead, being an evolutionary taxonomist, despises cladists. He may very well also dislike Marxists, but he need not: it only matters that others are apt to dislike them. By associating the two, Halstead hoped to suppress cladistics" (James S. Farris, "The Pattern of Cladistics," in *Cladistics* 1 (1986): 198.) For a delightful account of the classification of cladistics, see James R. Carpenter, "Cladistics of Cladists," in *Cladistics* 3 (1987): 363–375.

trenchant article in *Cladistics*, Cranston and Humphries criticize Seather, who, they say, argued that all "neocladists" (those who use computers in cladistics) and those who subscribe to the ideas of transformed cladistics contradict the views of Hennig. The basis of their criticism lies in defense of their own classifications of chironomid midges and the methods that once used to produce them. In order to contradict Seather, they undertook a computer analysis: "Cladistic analysis was undertaken using Swofford's (1985) package PAUP version 2.4 installed on a Cyber mainframe computer and version 2.4.1 on an Amstrad 1512 PC. The purpose of this comparison was to evaluate generalized additive coding methods for continuous data on 60-bit and 16-bit computers, respectively" (Cranston and Humphries, 1988, 76). So what we have here is a debate between two interpretations of a cladistic classification of chironomids being settled by the operation of a particular computer program on a particular machine; both specifications were important to the argument.

What the cladistics programs do is follow a given algorithm for generating and testing cladograms—phylogenetic trees, roughly speaking. In general, there is variation between programs in the weighting of desirable features such as the least amount of "convergent evolution"—independent development of a given characteristic in different lineages. Felsentein (1988) noted that computer methods had moved to the forefront in classification work:

> From small beginnings 10 years ago, the use of explicit algorithms for inferring phylogenies has grown to become the standard approach today. The acceptance of statistical approaches is increasing, and so has the availability of computer programs to carry out various methods Farris's FORTRAN program WAGNER-78 was the first phylogenetic program to achieve any currency. It has been supplanted by several larger and more sophisticated packages. My program package PHYLIP was first distributed in 1980, and has since been provided, free of charge, to over 700 installations. Farris's package PHYSYS was released in 1982 and is distributed for a substantial charge. Swofford's widely-distributed program PAUP was released in 1983, and is inexpensive. A particularly interesting program is Maddison's MacClade, a program for the Apple MacIntosh which allows the user to interactively manipulate the tree and watch the resulting changes in the distribution of character state changes. (119)

The problem here is that however fast the computer, the packages produce variable results and cannot possibly look at all the possibilities (and so an algorithm must be introduced to cut down the number of possible searches, and each algorithm will contain its own biases). Here is Yeates's recent estimate (1995) of the computer power needed to produce a complete history of the earth's species:

> Chase et al. (1993) conducted one of the most ambitious simultaneous cladistic analyses of many terminals, with a data matrix of 500 species, including all major groups

> of seed plants.[3] The authors noted that their tree searches took up to 4 weeks to complete, and the sets of shortest trees filled available computer RAM. An undiscovered set of trees of the same length exists. It is clear that even the fastest computers will only enable us to include a tiny fraction of the form in many higher groups as terminals in a simultaneous cladistic analysis using current algorithms. An analysis including all 10–30 million species currently estimated on the globe . . . would represent an astronomical number of dichotomous resolutions . . . and it would take a considerable number of researcher-lifetimes to complete. (344)

The reason this is so difficult is that it is an NP-complete problem, meaning that there is such an exponential increase in the number of possible solutions as the number of species increases that computers will always be outstripped past a certain point. The past is just as disordered in our databases as it was in planet earth in Lyell's day. . . . Indeed there have been moves to go back to simpler definitions of species (Mayr 1988) or to use entirely artificial, local systems (Hawksworth and Bisby 1988) so as to produce a pragmatic solution to the difficulties of the system. (Currently in fashion are GUIDs—globally unique identifiers—borrowed from industry, as is the metaphor of "barcoding" life).

This database-driven approach to the history of life is complemented by extended uses of the metaphor of computers and their programs in the historical sciences. Life is described as itself a program, with DNA being code. At the limit, if everything is information, then life can equally well be "stored" in computers as on the face of the earth. Specimens of what there is and has been on the face of the earth are very difficult to maintain; they need special conditions of humidity and temperature, and they take up a lot of space. As Curator Markus (1993) noted in 1993: "Some major collections . . . have small or inadequate staffs . . . Some botanists at the University of Queensland have suggested that we not keep all of the herbaria specimens now maintained as vouchers for ecological studies, as well as the large numbers or replicate specimens or series. They suggest that we rely instead on the published documentation of species. This would economize the maintenance of collections and increase the resources available for research" (37). Cutting down the number of collections is a general move in the information sciences. The general problem is that books and papers are in some ways very inefficient means of information storage: they take up a lot of space, and are very difficult to manipulate. We get right back to the archives embedded in the earth itself with the suggestion that the difficulties of maintaining biodiversity out there (in an

3. This was a large-scale analysis of the phylogenetics of seed plants, using sequences from plastic gene rcbl.

ever-shrinking "nature") can be solved by keeping a computer record of the genetic makeup of various plants and animals, after which they can be recreated on an as-needed basis. This merger of the archive in the computer with the archive out there is a final token of the buckling together of the problems of reading the archive that Lyell noted with the problems of maintaining stable manufactured archives.

The pervasive metaphor of the computer supports the application of database technologies to the biological record through prioritizing the use of information science techniques; and the biological record thus sorted lends weight to the computer metaphor as organizing principle—each bootstraps the other. There is a radical discontinuity between the two cases given here—the 1830s and the 1980s and 1990s; they are based on incommensurable theories of information and its technology.

As discussed in the introduction, Derrida (1996) argues in *Archive Fever* that with the development of computing technologies, we are creating new kinds of traces—new kinds of archives that give rise to new kinds of past. He typically does not carry this argument through to the natural sciences. We are in the process of creating a highly complex set of tools for retaining and reordering the past. There has in general been a technologically driven revolution in the historical sciences in the past thirty years—to such an extent that Lyell's uniformitarian reading of the history of the earth leaving no vestige of its own creation has been completely overturned (Allègre 1992). The proportions of the chemicals in the earth are used to tell an astronomical history, ratios of some radioactive elements allow the earth's surface to be traced back beyond a "complete revolution" through the core. I have argued that this new historicism is technologically informed, and that that technology is just as fraught a record keeper as the roving commissioners that Lyell so roundly roasted.

Why does the use of technology and technological analogy matter in the construction of an archive of history of the earth? It is clear that there is no way such an archive can be constructed without technological mediation. Where the interest lies is in the question of whether the nature of the technology used either directly or analogically produces secular changes in our representation and understanding of the past.

The secular change marked in the two main sets of examples I discussed can be summarized fairly briefly. We are going from one industrial and bureaucratic era (steamboats, trains, clocks, statistical thinking) to another (computers, the information age, database technologies). At the same time, we are going from one technological ideal (the world is basically *like* a clock—all the apparent evidence of confusion and catastrophe notwithstanding) to another (the world is basically *like* a computer with life as a series of programs—all the apparent evidence of contingent historical developments notwithstanding). So the ideal reading of the archive (the clock modeling the time of the printed page; the program modeling the sequentiality of computer instructions) is also both the ideal form of the archive (a strict and regular chronology in the nineteenth century, a generative algorithm in the twentieth) and its material base (paper files organized by epoch within linear time and a determinate past; computer files organized by multifaceted classifications and with an infinitely reconfigurable past).

Now the fact that we are now dealing with reconfigurable pasts—whereby new cladograms can be constructed on the fly with new classifications resulting—can be seen as simple progress. It could be argued that if they could have done it like that in the nineteenth century, then they would have; it is just that we have only in this century developed the information-processing capacity to produce a more useful archive of the earth's history. On the contrary, I maintain that we do not as a society have a series of separate and separable discourses about the past, each of which has its own problematic and development. Rather, I would argue, the boundaries between the disciplines are porous precisely because the same kind of information-processing technology is being used in each case (and in turn, the compelling logic of the change of information-processing technology is precisely what allows the various disciplines to appear to have their own very separate histories, since the same series of steps are taken logically *within* each specific discipline).

This reference to the significance of the underlying information technology is not particularly a deterministic or reductionist one: the development of that technology is based itself on a noncentralized but philosophically highly rich discourse about the nature of record keeping—and that development can be sparked equally by social forces or scientific developments. Information technology, defined as we have as both the material base (e.g., computers) and its ideal machinery (e.g., computer programs) is a contingent historical locus at which scientific and social discourses meet and can share skills, insights, and tools. Thus the technology does not have logical preference in driving the sciences (or society), and inversely these latter do not drive the technology. An ecology of technological memory practices within the historical sciences is a key site for understanding their framing of the past. Their technical substrate links strongly with their interior development, to give at any one epoch a picture of inevitability; and between epochs a set of radical discontinuities (Veyne 1971).

4

The Mnemonic Deep: The Importance of an Unruly Past

Presumably man's spirit should be elevated if he can better review his shady past and analyze more completely and objectively his present problems. He has built a civilization so complex that he needs to mechanize his record more fully if he is to push his experiment to its logical conclusion and not merely become bogged down part way there by overtaxing his limited memory. His excursion may be more enjoyable if he can reacquire the privilege of forgetting the manifold things he does not need to have immediately at hand, with some assurance that he can find them again if they prove important.

—Vannevar Bush, "As We May Think"

Forty-eight pitiable coevals, the schoolmistress on the right, on the left the myopic chaplain and on the reverse of the spotted grey cardboard mount the words "in the future death lies at our feet," one of those obscure oracular sayings one never forgets.

—William G. Sebald, *After Nature*

Introduction

When we want to remember something, we give it a name. An ineffable smell is harder to capture in the archive than a Latin binomial—especially since the Latin binomials were invented precisely as memory devices for field biologists. Within the field of biodiversity, one of the great questions is what kinds of things we should be naming so that we can keep the historical development of the planet "in mind" as we manage its future. Do we need to know about past communities, species, demes, ecologies, topographies, landscapes, soils? If so, how do we tease apart separate storable information bytes out of the flow of events or landscapes? We now explore how naming practices within the field of biodiversity have complex historiographies attached to them.

Epigraphs: Vennevar Bush, "As We May Think," *Atlantic Monthly* 176, no. 1 (1945): 108; William G. Sebald, *After Nature* (London: Hamish Hamilton, 2002), pp. 84–85.

Biodiversity has very quickly, from its coining in the 1980s (Takacs 1996; Wilson et al. 1988) become one of the keywords for our generation—recognizable in popular literature, in political circles, and in a myriad of scientific disciplines even though it has not yet graduated to the pages of the *Oxford English Dictionary.* There is general agreement in environmental circles that biodiversity is a good thing, but very hard to define. For example, there is significant cleavage between views of biodiversity as being about the number of species in a given locale (more species = more biodiversity) or about the amount of genetic information being held (a single species holding an isolated position on the "tree of life" can contain much more information than several variants species of a more common genus) (Eldredge 1992). This cleavage has policy implications: should one protect, say, a given wetland with one such isolated species but few others over another with many species in danger of extinction but that all have close relatives elsewhere (Faith 1994)? It also reveals deep differences between scientific communities: for example, it speaks to the different working practices and analytic tools of molecular biologists and ecologists.

Since a wide array of sciences contributes to building knowledge about the nature and protection of biodiversity, scientists and information professionals have been working to work out ways of sharing knowledge between them. The problem of sharing data across disciplinary, institutional, and national boundaries is a nontrivial one—for example, there are issues of who takes responsibility for maintaining databases; of the ownership of data (should access be free, even when it can lead to a commercial discovery?); and of the degree of context that must be provided for scientists using the database (who may not be aware of the conventions and practices of the scientists who produced the data set.

One overarching rubric covering these issues is that of metadata: "data about data" (Dempsey and Heery 1998). The problem I turn to now is what kinds of issues are encountered when moving between the precise conceptions of metadata that have been formulated in the literature, and the messy daily practice of doing biodiversity-related work. The best comprehensive definition of metadata is given by Michener et al. (1997). Drawing on the model of the highly successful Federal Geographic Data Committee (FGDC) for geographical data, they suggest that a good database in ecological science would include the following classes of metadata:

• *Class I.* Basic attributes of the data set (data set title, associated scientists, abstract, and key words). The purpose of Class 1 descriptors is to alert potential secondary users to the existence of data sets that fall within specific temporal, spatial, and thematic domains. In many cases, a short summary of the "data set usage history and question and comments from secondary users could be used to identify potential uses of the data and to highlight major strengths and weaknesses."

• *Class II.* All relevant metadata that describe the research leading to development of a particular data set. Two subcategories: description of the broader, more comprehensive projects—giving broader scientific context for an individual study; all pertinent information related to the research origins of a specific data set. Site characteristics, research methods, etc. Also permit history may be useful here; and also legal and organizational requirements that shaped the data.

• *Class III.* The status of the data set and associated metadata—copyright, proprietary restrictions, etc.

• *Class IV.* All attributes related to the structure of the data file—all variables should be labeled, defined, and characterized.

• *Class V.* Includes all other related information that may be necessary for facilitating secondary usage, publishing the data set or supporting an audit. May need to go back to original physical records for an audit, or to programs, etc. (Michener et al. 1997, 332–333)

There is little doubt that data wrapped in such a complex set of layers would be useful over the long term. It is also clear that it is very difficult to work out a path for getting from the data collection and storage methods currently being used in biodiversity research to a universal set of formal descriptors of data structures. In figure 4.1, we see the state of much of the available data. It

Figure 4.1
Hu cards, used in describing the flora of China . . . scanned in but not accessible. *Source:* Missouri Botanical Gardens.

represents one of upwards of 300,000 cards currently being scanned in by the Center for Botanical Informatics at the Missouri Botanical Gardens. The cards represent collectively many years of research by Dr. Hu, and they are an invaluable resource on the flora of China. However, they cannot easily be conjured into any of the five classes of metadata that Michener et al. suggest: each card would have to be processed individually, and even then several data fields would be left empty in any electronic representation of the database. Even if, in the best of all worlds, the metadata were created and the data on the cards were conjured into searchable form, it is likely that some contextual information—such as, for example, what was crossed out and what was added in later—would be missed.

The issue of what happens to data as it gets from the field to the scientific article has been extensively explored in the science studies literature (Goodwin 1994; Latour 1993a). There has been little analysis of what happens as one gets from raw data to databases (and even less of the move from databases to analysis)—and whether decisions taken in this process have continuing effects on the interpretation and use of the resultant data stores. Issues of how data is worked into storable form are integrally issues about how groups of scientists communicate with each other and with governmental and other agencies. This means that the development of metadata standards should go hand in hand with an exploration of modes of scientific communication—currently, the two areas are effectively disjunct. I will argue further that decisions taken at the level of data standards (memory practices) very quickly become issues of historiographical import within the sciences they concern. It is extremely difficult to tell stories that cut across one's data structures: one can't in general find the information, and even when one can it is often not in usable form. I cover now, in turn, issues of naming, retaining context, and integrating information.

Things that Are Hard to Classify

There are many things, as Douglas Adams (Adams and Lloyd 1990) reminds us in *The Deeper Meaning of Liff*, that we would like to talk about but just don't have the words for.[1] In general, you cannot develop a database without having some means of putting data into pigeonholes of some kind or another; you can't store data without a classification system. (I say "in general" because there are systems that claim to do classification work after data entry—such as

1. For example, "naugetuck," or "a plastic packet containing shampoo, mustard etc., which is impossible to open except by biting off the corners" (Douglas Adams and J. Lloyd, *The Deeper Meaning of Liff* (New York: Harmony Books, 1990), p. 72).

systems that produce automatic, dynamic classifications of large bodies of free text data. For our purposes, though, the problem only recurs with these systems, since the free text itself needs to have some kind of regular naming system—or set of naming systems—for it to be amenable to useful automatic classification).

A major problem in the world of biodiversity is that while classification systems are about kinds of things (flora, fauna, communities, etc.), the world of biodiversity data is radically singular. Just as species can be endemic to very small areas, so too can data about species—museums, a thirty-foot strip along the side of roads, and libraries are where most research is done. As Heywood et al. (1995) write: "It has to be remembered that the vast majority of species described in the literature are 'herbarium' or 'museum' species and their existence as coherent, repeatable population-based phenomena is only suppositional" (48). Raven et al. (1971) remark that this throws into relief the fact that our names for organisms do not contain much information: "The taxonomic system we use appears to communicate a great deal about the organisms being discussed, whereas in fact it communicates only a little. Since, in the vast majority of instances, only the describer has seen the named organism, no one with whom he is communicating shares his understanding of it" (1212). There is often only one of a given species in preserved form, so that in order to check one's own samples one needs an exhaustive global database (in the form of a robust body of literature—more on this topic later—or in searchable electronic form) and a means of transporting type specimens from one site to another. So working out what is in one's own collection as a prelude to cataloguing it and putting it into searchable form represents a bootstrapping problem: unless you have described your collection well, others can't describe theirs; but equally you can't describe yours until they have theirs. In a study of the International Classification of Diseases, Bowker and Star (1999) remarked that this bootstrapping problem was a common feature of the development of global databases.

There are many entities or communities that, in principle, one would like to be able to classify in order to get good comparative data, but that do not prove easily amenable to classification—and so have led to the proliferation of local data stores or to no data collections at all. This sets up a reverse bootstrapping process, whereby things that cannot be described easily and well get ignored, and so receive an ever-decreasing amount of attention. This has been a problem for the fossil community, for example, the first class of things that are hard to classify. Fossils are preserved in several different fashions, and each mode of preservation favors its own set of body parts (Galtier 1986): "The number of currently recognized compression species is much too high due to underestimation of individual variation and poor knowledge of the leaf polymorphism. Many fewer species have been distinguished in permineralized material" (12)—meaning that it is often very unclear whether one is dealing with a new species descended from one below it (following the general rule that going down means going back in time), or whether one is dealing with a variant leaf form of the same species. It is often not clear how to match parts of plants (leaves and stems and flowers), let alone seeds recovered from a midden with any given leaf (compare the descriptions of the various historical reconfigurations of body parts—and the successive, wildly diverging illustrations of the biota—in the Burgess Shale by Gould (1989). However, fossil information is extremely valuable in, for example, the mapping of oil fields; you can track down unconformities in the subsoil through characterization of fossil communities. N. F. Hughes (1991) notes the reverse bootstrapping:

> The use of fossils has in recent years come to be regarded as cumbersome and unproductive; the work is said to abound with unimaginative complacency, with the obscurity of esoteric terminology, and with lack of compatibility of treatment of different groups including even that between the fossils of plants and animals. As a result, much effort has been directed by geologists in solving their stratal problems towards employing any other available physical or chemical phenomena, and thus to avoiding altogether the 'expensive' and supposedly ineffective use of palaeontologists and their fossils.
>
> Palaeontologists, who almost all continue to believe that their fossils and the distribution of these form the only viable method of discriminating diverse and confusing strata, are striving to present their fossils more ingeniously and to win back the confidence of the geologists. (39)

His response has been to produce a "paleontologic data handling code" that does not force one to choose a taxon descriptor for a given fossil. It is based, he claims:

> first on equal treatment of all observational records, to avoid the loss of primary information by burial of records within taxa, which for fossils are inevitably subjective, and which render the original records uncheckable. . . . All observational records are made of paleotaxa (of similar scope to species but immutable) or as comparison records formalised to require an instant author decision on the degree of similarity. As the traditional generic name of fossils conveys no precise information, a binomial is completed instead by means of a 'timeslot' name based always on a timescale division of the Global Stratigraphic Scale of the International Union of Geological Sciences. (ibid., 42)

This solution in turn leads to its own problems—first, with respect to ensuring adoption (no easy matter since there are numerous groups using fossil data, often with their own well-entrenched schemata, and second, with respect to the stability of the global stratigraphic scale and of dating techniques: we will return to this later. This has led to the situation where geologists from some petroleum companies have abandoned the scientific literature and developed their own coding for fossil remains. Hawksworth and Bisby (1988, 16) describe the use of "arbitrary coding systems" by some prospectors. And even when the agreed-upon naming procedure is followed, the fossil genera might be buried in, say, the *Journal of the Geological Society of London* (Boulter et al. 1991, 238), where no database manager or neobotanist would look for a new genus. As one author notes, there are just too many conflicting uses for fossils: "Many of the difficulties in palaeobotany and its sub-discipline, palynology, occur as a result of the sometimes conflicting aims of botanical and geological researchers handling these data" (ibid., 232). The point here is that the bootstrapping work of getting any classification system off the ground is, as Hughes notes, always a battle of "winning confidence" and "ingenious presentation" and that without this work, no technical fix to the data storage problem will ever get established.

But it's not just the world of data that is radically singular; so is the world itself. A second class of things that are hard to classify are soils. In an article discussing "the monumental (or dinosauric) *Soil Classification, A Comprehensive System—7th Approximation*," published in 1960 by the Soil Survey Staff of the U.S. Department of Agriculture, Bennison Gray (1980)) discusses the problems of classifying something that does not break up into natural units. Different sets of researchers will have very different intuitive definitions of what soil is. For many agriculturists, soil is just something you can or cannot grow crops in; for ecologists, soil can include hard rock (to which lichen can cling) and be defined as that part of the surface of the earth subject to weathering influences—a comprehensive but itself imprecise definition. Soil, he notes, is three-dimensional—but there it is an indexical question how far down or across you go to determine a unit (or pedon): in the prairie states of the United States, there is a lot of uniformity in the topsoil; in California with its complex geological history, much less so. Thus the ped/pedon/polypedon nesting will be different in different regions—the ped being a crumb of the soil; the pedon being "the smallest volume that can be called 'a soil'" and designated by the "vague and somewhat arbitrary limit between soil and 'not soil'"; and the polypedon, defined by U.S. Soil Survey Staff, "a cluster of contiguous pedons, all falling within the defined range of a single soil series. It is a real, physical soil body, limited by 'not soil' or by pedons of unlike character in respect to criteria used to define series" (ibid., 139–140). There are major national variations in classification in the United States, Australia, Belgium, Canada, Holland, England, France, Germany, New Zealand, Russia, and Scandinavia:

"And, because none of these countries are tropical, soils of the tropics are not given detailed coverage in any of the nationally oriented taxonomies" (ibid., 141). Huggett (1997) remarks that "soil classification schemes are multifarious, nationalistic and use confusingly different nomenclature. Old systems were based on geography and genesis. . . . Newer systems give more emphasis to measurable soil properties that either reflect the genesis of the soil or else affect its evolution" (178). The local variations in classification are a variant on a major theme in geological history: because each region of the earth has its own complex story written into its soil and subsoil, a "genetic" classification (which has been, since the mid- to late nineteenth century, the canonical form of universal classifications in many fields; see Tort 1989) cannot so much uncover types as the singularity of every story.

Along the same vein as soils, landscape topology has proven very difficult to classify. Huggett (1997) notes that "many landform classifications are based purely on topographic form, and ignore geomorphic process" (228)—in our terms, they are not genetic but descriptive. He notes that the Davisian "geographical cycle" in the late nineteenth century was the first modern theory of landscape evolution:

> It assumed that uplift takes place quickly. The raw topography is then gradually worn down by exogenic processes, without further complications from tectonic movements. Furthermore, slopes within landscapes decline through time (though few field studies have substantiated this claim). So, topography is reduced, little by little, to an extensive flat region close to base level—a peneplain. The reduction process creates a time sequence of landforms that progresses though the stages of youth, maturity, and old age. However, these terms, borrowed from biology, are misleading and much censured (ibid., 230).

In more recent thinking, steady states in the landscape are the exception rather than the rule (ibid., 236), and indeed there is an argument that landscapes themselves are evolving over time; evolutionary geomorphologists argue that rather than an " 'endless' progression of erosion cycles," there have been several geomorphological revolutions (ibid., 258). This ties in directly to biodiversity arguments, since the argument has been made that there is a growing geodiversity subtending our current relative biodiversity peak—itself offset by the current extinction wave.

A third class of things that are hard to classify are things that don't have easy boundaries and are of indeterminate theoretic status—for example, the concept of communities in ecology. In the first half of the twentieth century, Clements and Gleason took opposing views on the existence of natural communities (Journet 1991). Clements said that there were natural large units of vegetation, such as a forest. He saw the community as a "superorganism" (ibid.,

452). These communities go through predictable stages of growth and development—called "series." There are natural climax communities optimally suited to a given (stable) climate regimen. Gleason saw the groups of plants as being less an organism than a coincidence: "the implication of Gleason's arguments is that what an ecologists calls a community may be only an artificial boundary drawn around a collection of individuals" (ibid., 455). He rejected the overall stability of the climate. In the new catastrophism (Ager 1993), there are many who have come to accept the position that stability, the prerequisite for many definitions of communities, just does not obtain sufficiently to make this a useful concept. However, some of the more exciting work in biodiversity research, GAP analysis, relies on a pragmatic definition of community, which includes many ad hoc assumptions, recognized as such by its developers (Edwards et al. 1995). GAP analysis involves taking aerial photographs of a given area, then marking off natural vegetation communities. Vegetation communities are then used as surrogates for animal communities. One of the difficulties with this form of analysis is that aerial photographs can only be used to characterize vegetation communities with about 73 percent success; vegetation communities are difficult to define anyway. Even if they are successfully defined and located, they are not necessarily good surrogates for animal communities (free-ranging predators might not distinguish between any two vegetation communities)—bats with point distribution (in caves) cannot be picked out; in general, structural features of the landscape are not mapped, so that if a species requires a given type of setting within a floristic community, the setting has to be assumed to exist (ibid., 4–2).

So we have covered here three kinds of things that are hard to classify: entities where the data itself is singular and scattered, entities that are not amenable to genetic classifications and are singular, and entities that may or may not exist. If we take the first two cases, the first generalization to emerge is the relatively obvious one that when either data or the world is singular, then naming schemes tend to break down. More interesting is a corollary—that once they do break down, these things do not get represented in databases, or if represented are represented in incompatible forms across different sites. This in turn means that less attention is paid to such entities. If we take the last two cases, the obvious generalization is that things amenable to genetic classifications against a stable temporal backdrop can be relatively easily named, but when time itself is a dynamic variable then the naming mechanisms tend to break down. This can be combined with the first generalization in the argument that things that are singular either in space (one unit of space is not equivalent to another) or in time (the history of the earth is too secular) cannot be easily named. As I write this, I am aware that this is on one reading a remarkably obvious statement. It is hard to find generic names for singulars. Where it takes on depth is that it takes us very quickly from data structures into constraints on the historiography emerging from the sciences of biodiversity. If certain kinds of entities are being excluded from entering into the databases we are creating, and if those entities share the feature that they are singular in space and time, then we are producing a set of models of the world that—despite its avowed historicity—is constraining us generally to converge on descriptions of the world in terms of repeatable entities: not because the world is so but because this is the nature of our manipulable data structures.

Things that Do Not Get Classified

We can develop this point further by looking at things that do not get classified. There are certain kinds of plants, animals, and systems that are charismatic. These in turn create a set of others, entities that are often just as important, but that are overlooked because they do not lead to spectacular science or good funding opportunities.

The use of the word *charismatic* here is adopted immediately from the biodiversity literature, though the concept itself traces back to Weberian sociology: for example, Edwards et al. (1995) talk about the problem of getting too much information about "charismatic megafauna" (1–1). Certain species are more likely to get attention from policymakers and the public than others. Many more care about the fate of the cuddly panda, the fierce tiger, or indeed the frequently drunk and scratchy koala bear than about the fate of a given species of seaweed (or sea vegetable, to use the more recent, kinder coinage). And this attention has very direct consequences. On the one hand, scientists are more likely to get funding for studying and working out ways of protecting these charismatic species than others; and on the other, people are more likely to become scientists with a view to studying such entities—another feedback loop that skews our knowledge of the world.

Furthermore, they are more likely to get funding for using exotic, expensive tools, as a participant at a symposium in 1988 noted: "Something called Stearn's Law has entered the literature, following a remark of mine at a conference in 1968. This states that at a given time the perceived taxonomic value of a character is directly related to the cost of the apparatus and the difficulty of using it" (Hawksworth 1991, 176). Readers of *Science in Action* (Latour 1987) will find this no surprise. An unfortunate corollary that is central to current nonnaming problems within biodiversity is that taxonomic and systematic work, which particularly in the morphological tradition uses noncharismatic technology, has consistently lost out to more "exciting" areas of research that do not try to provide consistent names—that necessary prerequisite to good biodiversity policy. Rendering more stark this picture of things that do not get classified is the problem of the "disappearing taxonomist." Thus in the Taxacom Discussion Listserv, a listserv devoted to questions of systematics, a thread was opened in January 1999 on this topic: "besides the self-regret and crying, how to be a taxonomist in a primarily non-taxonomic world" (Taxacom Discussion Listserv, 1/29/99). Wheeler and Cracraft (1997) complain that "faculty positions in systematics have been replaced by non-taxonomic, often molecular, ones" (444); echoing remarks from the 1950s (Vernon 1993), Gunn et al. (1991) bemoan "the increasing migration of taxonomists away from the herbarium and into the laboratory" (19). This is paradoxical indeed, since as Hawksworth and Bisby (1988) remark after deploring the shortage of taxonomists: "The demand for systematics is high and increasing in conservation, environmental monitoring, agriculture and allied sciences, biotechnology, geological prospecting, and education and training. The requirements of consumers for stability in names, reference frameworks, relevant taxonomic concepts, user friendly products, and communication need to be addressed by systematists if they are to meet this demand" (1). So, on the one hand, we have increasing demands for stable classification systems in order to meet the data demands of a burgeoning set of scientists and entrepreneurs, and, on the other hand, we have taxonomy[2] as low-status work not attractive to incoming students. There have been attempts to solve this problem in the past ten years—such as the NSF PEET (Partnership for Expanding Expertise in Taxonomy) or the discussion of the use of parataxonomists in biodiversity data collection, their skills being augmented technically with the development of interactive "keys" for plant identification available on handheld devices. However, the general problem remains that across the sciences, the activity of naming is mundane and low status, even though it is an activity central to the development of good databases. Insofar is it remains—or becomes even more—so, there is little prospect of an abundance of taxonomists describing the plenum of nature regardless of the relative status of problems within taxonomy. To the contrary, the skewing of our data collection and management efforts promises to continue unabated.

2. I will use the words *taxonomy* and *systematics* in accord with the following definition of the products of taxonomy being: "a taxonomic information system comprising classification, nomenclature, descriptions, and identification aids. Systematics is then taxonomy *plus* the biological interrelations—breeding systems and genetics, phylogeny and evolutionary processes, biogeorgraphy, and synecology (partricipation in communities)" (D. L. Hawksworth and F. A. Bisby, "Systematics: The Keystone of Biology," in *Prospects in Systematics,* edited by D. L. Hawksworth (Oxford: Clarendon Press and Oxford University Press, 1988), p. 9).

We are, then, dealing with a finite group of taxonomists trying to generate a world of data. Heywood (1997) characterizes those species that tend to attract most-favored-entity status: "our knowledge of biodiversity in terms of named organisms is highly skewed in favour of certain groups such as birds, mammals and flowering plants, rather than invertebrates, fungi and microorgamisms, except when the latter are of vital importance to humankind in terms of economics or health" (9). So, in general, we tend to protect and study things that are about our size or larger and things that are spectacular in one way or another. As far as biodiversity goes, microscopic beings are doing very well indeed—there are some archaea (prokaryotic single-celled organisms from the depths of time) in the urine-soaked walls of Paris, for example. In the plant kingdom, vascular plants get the nod, as Klemm (1990) notes: "An analysis of plant protection legislation in 29 European countries shows that in most cases the lists of protected taxa are relatively short, seldom exceeding one hundred entries, are almost always entirely composed of vascular plants and are largely dominated by spectacular species which are attractive to collectors to the public" (30). Very small things often get the least attention—Rita Colwell (1997, 282) points out that there are more viruses than bacteria in the open ocean. These viruses do not receive scientific attention commensurate with their numbers (for a proudly viricentric view of the world—where many of our arrows of biological causation break apart—see Hurst 2000). When the exotic meets the mundane, the former often wins out. Judith Winston (1992) tells the sad story that she has found it much easier to get funding to study Antarctic bryozoans (sessile moss animals that form colonies) than those in local coastal waters off the northeastern United States. She says that this skewing continues to this day, such that "many of the most urbanized coastal areas, the very areas in which human impact has been greatest, have never been surveyed. The practice continues today—if systematics has any glamour at all, it is only when carried out in the most exotic locale" (ibid., 157). In a related problematic choice, the Long Term Ecological Research Network in the United States and Antarctica devotes significantly more resources to "pure" than to "transformed" landscapes—even though the latter predominate. So what gets studied is the exotic other. And if the choice is between two others, then the more exotic will be chosen. Tim Boyle and Jill Lenne (1997) note that small and unusual cropping systems, like home gardens in Java, have been intensively studied, whereas very little is known about "the vast millet, sorghum, cowpea and groundnut subsistence systems of sub-Saharan Africa or the slash-burn systems of South and South-east Asia and South America" (33).

Terry Erwin (1997), famous for his early estimates of the global numbers of species, notes a kind of feedback loop common to data collection efforts[3]:

> The *described* species of beetles, about 400,000+ (Hammond, 1992), comprise about 25% of all described species on Earth. This dominance of beetle taxa . . . in the literature has resulted in Coleoptera being perceived as Earth's most speciose taxon. Thus, it has garnered further taxonomic attention from young taxonomists which in turn has resulted in more species of beetles being described than in other groups. Beetles are relatively easy to collect, prepare, and describe, significantly adding to their popularity. Such unevenness in taxonomic effort may or may not give us a false picture of true relative insect diversities. (29)

Erwin makes two points here: first, that we tend to study what has already been studied, which is a version of the Matthew effect, and, second, that beetles get studied because they are easy to study. Similarly, c-elegans was sequenced as a model for the sequencing of the human genome partly because it was very easy to deal with (it could be frozen, shipped off via post, and then reanimated). In figure 4.2 we see a graphic illustration of the connection between apparent species richness and intensity of study of fossils from various epochs.[4] It can, of course, be argued that there are more people studying the recent periods because there is more to study (more outcroppings—though not so much more) and larger area. However, it is clear overall that there is also a weighting of our global databases in favor of those entities that can be readily studied.

Things that people don't generally like or need get less attention—both in the real world and in naming. In botanical nomenclature, when plants are considered of particular value, then their names can be protected against the ravages of the inquiring taxonomist who finds a prior description of the species under another name. This is an information retrieval issue: protection is put in so that searches through the literature for such plants will not become unduly complicated. However, weeds—even when economically important—frequently do not get this protection, as Gunn et al. (1991) complain:

> In cases where names of economically important species are involved, the Code provides for conservation in order to preserve current usage. Judging from the success of

3. Hammond discusses a "species inventory" of beetles.

4. Compare Dumont on rotifers: "About 2000 'species' (by which I mean taxonomic bonomens or trinomens) of Rotifers are now known throughout the world; ca. 1350 of these occur in Europe (Berzins 1978). This is no coincidence, and relfects the distribution of rotiferologists (in Scandinavia, Germany, and Great Britain) rather than that of Rotifers" (H. J. Dumont, "Biogeography of Rotifers," *Hydrobiologia* 104 (1983): 20). Berzins gives a distribution of rotifers in North America from surveyed bogs.

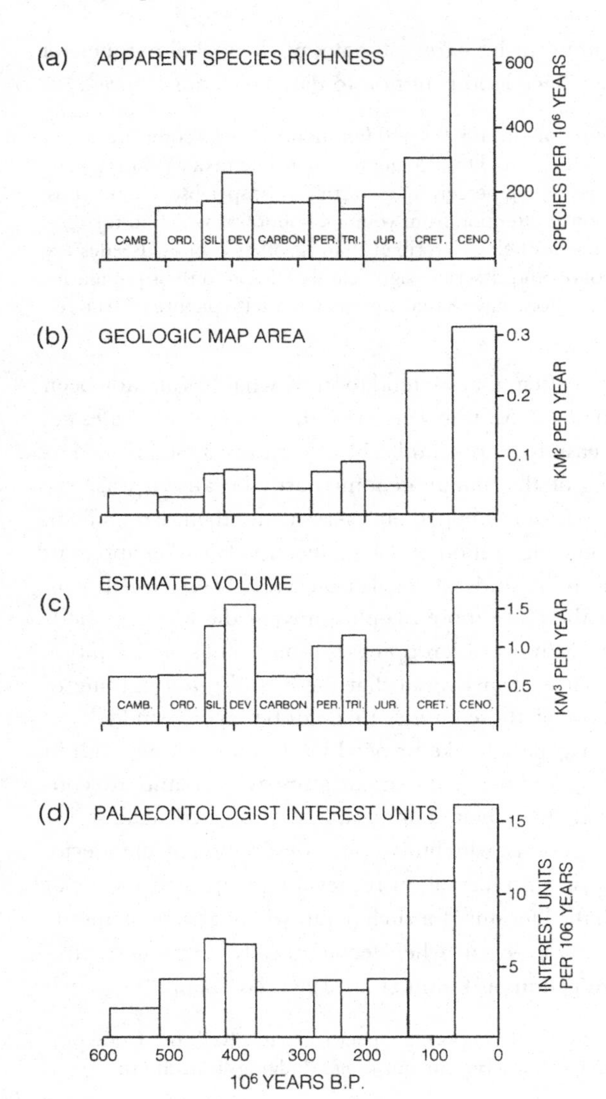

Figure 4.2
Apparent species richness. *Source:* P. J. Brenchley and D. A. T. Harper, *Palaeoecology: Ecosystems, Environments, and Evolution* (London: Chapman and Hall, 1998), 318.

past species conservation or rejection proposals, it is not always clear as to what constitutes an economically important species. Some weeds of significant agricultural importance have not qualified for name conservation. With the expanding potential of gene transfer for crop improvement involving more distantly related taxa, more species will become useful to agriculture. Communication about such species depends on a stable nomenclature. (18)

When entities have the misfortune to be small and generally disliked, then they will certainly not get the attention that others do. Such is the fate of the parasite. M. F. Claridge (1988) notes that the geographical isolation speciation concept, basic to much evolutionary theory, just does not apply to parasites, since for parasites speciation can very well occur within a given location. He notes that, in general, taxonomic literature has been anthropocentric, using the story of human evolution as a "reference point." The evidence for this, he notes, is the high frequency of human evolution being used as an example in textbooks and articles. The exclusion of the parasite has the kind of double effect noted previously; database tools are developed that can very easily process data that fits the geographical isolation speciation model, so you can swap techniques from one species to another. Conversely, little work is being done to represent data that represents other forms of speciation. The parasite gets harder and harder to represent as the new databases consolidate. At the other end of the production process, examples are chosen in textbooks that reinforce our world as one with the form of speciation particular to charismatic species.

Developing this latter point, Robert O'Hara (1992), in a paper about narrative representation and the study of evolutionary history, notes that "conclusions about evolutionary processes that are based on the structure of trees that have been, by selective simplification, brought into alignment with preexisting nomenclature probably say less about evolution than they do about the narrative character of the preexisting nomenclature" (O'Hara 1992, 153). He points out subtle factors affecting the representation of cladograms (trees representing hypotheses about phylogenetic descent or patterns of character traits among groups of species—depending on one's theory) such as the fact that you tend to choose a representation that includes "clades" (related groups of species) that have already been named, thus creating what he calls a "grooving effect" (ibid., 151).

I have defined these kinds of nonnaming processes elsewhere (Bowker 1994) as a form of *convergence*. What this means in this case is that a set of data structures and information retrieval models are set up so that a particular, skewed view of the world can be easily represented. With these structures and models in place, it is easier to get funding and support for research that reproduces this view—your work will be understood more easily, you can make good use of material from cognate areas, and so forth. Thus the world that is explored scientifically becomes more and more closely tied to the world that can be represented by one's theories and in one's databases: and this world is ever more readily recognized as the real world. Taking together the things that are hard to classify from the last section and the things that do not get classified from this one, we see the first threads of a pattern to the convergence that I have described.

One way of characterizing what we have seen in this section so far is a nested set of "centricities." First of all, any entity that is perceived as close to us in evolutionary terms or in terms of size, or any entity that is economically important, will receive more immediate attention in terms of taxonomic work, will be protected more eagerly and will be used for examples more often. Take, for example, vertebracentrism and anthropocentrism. However, we can follow the nested centricities further down into the laboratory—where what is interesting is what is in the interests of the scientist (commanding the largest set of resources—Stearn's Law, and the case of the bryozoans) or what is amenable to laboratory work (beetles) or what has interested other scientists. Together, these centricities entail a co-development of data structures and narratives that make humans indeed the measure of all things. The centricity pattern works immediately for all the examples given save one—the interest in the exotic other agricultural system. Here again, however, it can be seen along one axis, just as in the bryozoan case, that the exotic wins out over the mundane not in terms of information value but in terms of scientific cachet.

More fundamentally, the case of agricultural systems is also indicative of a set of issues around what can be learned from indigenous knowledge, and how information about indigenous action can be stored and used. Two points can be made. The first is that the question of whether pre-industrial action—fire management of the landscape as discussed by Pyne (1997) (see, e.g., Pyne 1991)—is "natural" or "cultural." For example, McIsaac and Brün (Brün 1999) argue that for some data models, the question of whether human action is endogenous or exogenous changes the subsequent ecological analysis. It has direct ecological implications as well, as Posey (1997) points out: " 'Wild' species are products of nature and, presumably, human societies have no special claim to them. Species or landscapes that have been molded or modified by human presence, however, are not automatically in the public domain and, consequently, local communities may claim special rights over them" (69). (This is the same logic whereby the World Bank assorts societies as "cultural" or "social"—the key difference being that you can't flood the territory of a cultural group). The question of whether (particular sets of) humans are part of nature, and so an input into a natural model, or beyond it is thus a philosophical/theological question, an economic issue, and a data storage concern. Second, it transpires that frequently even when so-called indigenous knowledge is being collected, it is not used: thus Cori Hayden (1998) describes how biodiversity prospectors fill in a data field about local knowledge since this is a regulatory requirement, but do not consult this field when testing a substance (they are convinced that their battery of tests is more exhaustive and efficient than local knowledge would be). The general tendency is to see pre-industrial humanity as "natural" and pre-industrial knowledge as hearsay—thus adding another layer to the centricities, one where industrial humanity is the axis around which data collection revolves. The problematic nature/culture divide is seen in the consistent non-classification of cultivars (cultivated varieties) bemoaned by Brandenberg (1991)[5]: "Anderson (1952) already put it in this way: 'Most modern taxonomists do next to nothing with cultivated plants; many deliberately avoid studying or even collecting them. As a result the scientific botanical name affixed to most cultivated plants becomes just an elaborate way of saying "I do no know"'" (24).

In general, what is not classified gets rendered invisible (Star 1991). Derrida (1980) wrote years ago about the importance of looking for the "other" categories—things that are bundled out of a given discourse as not significant. He argued that these "others" (what is excluded) often indicate the very problem that a totalizing philosophical discourse cannot deal with. Thus he wrote about the difficulty of speech act theorists in dealing with the category of humor, for example. The apparently heterogeneous set of others that we have uncovered in this section tells a consistent story. The negative telling is that things that do not get classified are not considered of economic, aesthetic, or philosophical importance—weeds, noncharismatic species, and indigenous knowledge in turn. The positive telling is that our databases provide a very good representation of our political economy broadly conceived: that which we can use through our current modes of interaction with nature and other cultures is well mirrored in our data structures. What gets excluded as the "other" is anything that does not support those modes of interaction. Put this way, it appears likely that there are shadow systems to the exclusions. It is surely likely that there are regularities to the ways of knowing and being that we fail to give a name to in the throes of our current "archive fever" (Derrida 1995). This is one of the ways in which the world converges (messily, partially) with its representation—that which can be represented is that which is counted,

5. A book summarizing the state of knowledge on plant speciation and evolution.

measured, protected, saved. As the representation becomes internally more manipulable, it becomes externally more apparently real.

Things that Get Classified in Multiple Ways

A banner headline in *The Independent* newspaper on November, 23, 1998, reads "Scientists Reclassify All Plants." When I first read it, I wondered why my own garden had not received its own scientist—clearly their bounteous energy did not extend to my back yard. And yet of course they did not actually reclassify all plants; reclassification is a long, slow, and expensive process, and there is no simple path from the molecular sequencing techniques referred to in the body of the article to the development of new plant classifications (see Group 1998 for the paper that led to the headline). Furthermore, even in the world of electronic databases we are moving into, there is no touch of a button that would allow us to bring in any new system. To the contrary, when a given database of plants, of the ecology of a given area, of paleontology, and so forth is designed, it necessarily draws on a contemporary classification—and will be very rarely updated (and very difficult to update) should the classification change. Even where a single set of names is adopted, there are problems of synonymy: the venerable *Index Kewensisi* has been estimated to offer 950,000 names for 250,000 flowering plant species (Lucas 1993, 11; cf. Hawksworth and Mibey 1957, 57, on synonymy among fungi). The result is a Tower of Babel, where numerous outdated classifications present themselves to the scientific researcher with equal force; indeed, they must be used if the associated data is to have any value. In order to explore the question of multiple classifications, I will look at the two- century-old effort to establish stable scientific names for plants, and I will argue that the attempts have been so difficult because precise boundaries for priority (who first named the plant), publication (where the name is published), and reach (who has the authority to name) are integrally questions about the organization of work in systematics and about the scientific features of a given plant. In so doing, I will draw attention to issues that have arisen with the need for long-term wide-scale information storage and retrieval (cf. Bowker and Star 1999, chap. 3) and will discuss the range of solutions that have been worked out over time—and how these affect the ordering of knowledge in the field of botany.

We will see a set of memory practices emerge around the issue of creating a stable global memory for science. Diana Crane (1972) claims that, in scientific literature, "The 'life' of a paper is very short, with the exception of a few classics. Papers published five years ago are 'old'. Papers published more than fifteen years ago are almost useless in many scientific fields" (8). Here I examine a field of science in which this is emphatically not the case—the field of botanical nomenclature—and discuss the issues that have arisen over the past 250 years as botanists have tried to develop universal, standard, consistent, stable names for plants (figure 4.3). The practice of botanical nomenclature is not fully aligned with the practice of botanical classification; although in principle Linnaeus's system is both classificatory and nomenclatural, in practice many names are retained beyond their classificatory currency. Crane's model works best in physics, where there is no assumption that information collected in the early nineteenth century will still be of interest to the current generation of field theorists. There is the assumption (e.g., Poincaré 1905) that new theories will reorder knowledge in the domain effectively and efficiently; and since Kuhn (1970) most would accept that a major paradigm change in, say, the understanding of "gravity" renders previous work on inclined planes literally incommensurable—not to mention technical improvements making the older work too imprecise. Chemists have a somewhat greater need to delve into older material (but certainly not where it concerns naming chemicals, since the procedure has been internationally standardized since the mid-nineteenth century). Astronomers trawl back further in time, seeking traces of supernovae in ancient manuscripts—but sporadically; they are just as likely to look at monastery records as at Tycho Brahe's original data.

In order to name plants, botanical taxonomists consistently need regular reference to scientific literature dating back to the mid-eighteenth century. Botanical nomenclature of vascular plants (those with conductive tissues) dates back to the first edition of Linnaeus's code in 1753. Since then, there has been a series of attempts to deal with the problem of name stability: how to ensure that a given species will have the same name all over the world and over time. In the 1820s the Kew Rule was developed out of Kew Gardens to deal with priority in generic names; in the 1860s, George Bentham worked on two large projects to stabilize systematics (Stevens 1997). In the 1890s the Berlin rule (limiting priority for names that had fallen into disuse or never been accepted) came into conflict with the Philadelphia rule (according to which priority was absolute). Throughout the past century there have been a series of international conferences to deal with the nomenclatural issues arising.

The principal problems are twofold. On the one hand, it is highly desirable to be able to change the names of plants when new scientific insights are developed. On the other hand, it is extremely difficult and costly to change these names. Consider the common tomato. Recent systematics research has suggested that its genus *Lycopersicon* should lose its status and be accreted once more into the genus *Solanum*. At the time a recent book on nomenclatural stability (Hawksworth 1991) was produced, the International Seed Trade Association was protecting the old name until the systematics debates were

Scientists reclassify all plants

BY MICHAEL MCCARTHY
Environment Correspondent

THE SCIENCE of botany has been turned upside down by a new classification of the world's flowering plants and trees based on their DNA rather than their appearance.

Worked out by a team led by scientists from the Royal Botanic Gardens at Kew, south west London, it has caused a complete rethink of the relationships between many plant families. It shows, for example, that the closest relative of the lotus, the sacred flower of Buddhism, is not the water lily it so much resembles but the the smog-resistant plane tree of London's squares.

The lotus (above) is not related to the lily but to the plane tree (top)

When it is published next month in the *Annals of the Missouri Botanical Garden*, the classification, which for the first time establishes the relationships of all plant families through their genetic material, will do away with 200 years of previous plant taxonomy dating back to Linnaeus. This has hitherto been based on flowers' and trees' morphology – their appearance and visual characteristics, such as how many leaves or petals they have.

But the ability to examine plants at the molecular level, which has become available on a large scale only in the 1990s, makes clear that many of the relationships botanists previously assumed from morphology are wrong.

There are many surprises in the new classification. The papaya is not related to the passion flower, as was previously thought – its closest relative is the cabbage family; roses are closely related to blackthorns, nettles and figs; and peonies are not related to buttercups but to the saxifrage family.

The classification is the work of nearly 100 scientists led by Mark Chase, head of the molecular systematic section of Kew's Jodrell laboratory, and two colleagues, Kare Bremer of the University of Uppsala in Sweden and Peter Stevens of Harvard. It has taken more than seven years and involved the detailed comparison of three genes for each of 565 representatives of all the families of flowering plants. Most of the work has been done at Kew.

"I think you would have to say it is a breakthrough," Dr Chase said. "There has never been such a focused effort at sorting out a major group of organisms as has been done here with the flowering plants, which are of course critical for life on earth, so its importance is economic as well as scientific."

Botany upturned, page 3

Figure 4.3
The work of the angiosperm classification group in *The Independent* (November 23, 1998).

completed—but within the Botanical Code (e.g., affecting scientific publications) the name could change during such discussion (Brandenburg 1991, 25). Name changes introduced as taxonomic theory develops can have large-scale economic consequences: it could cost tens of millions of dollars to relabel packets of tomato seeds, revisit regulations, and so forth (one commentator noted that "single name changes can cost the horticultural trade millions of dollars, and . . . nurserymen would go out of business if they took the matter seriously" (ibid., 1991, 30)). Brandenberg asks: "Have you ever tried to explain to a nurseryman, a plant trader, or a customs officer which name he should use for the tomato? Can you imagine the reaction if such a name has to be changed three times in ten years? If you have done so, you have a perfect explanation for the unpopularity of plant taxonomy amongst those in any work related to agriculture, horticulture, and silviculture: the inability of plant taxonomy itself to settle discussions with only a nomenclatural background" (ibid., 24). Anderson's account (1991) of the same problem raises the problem that even if a decision is made, the confusion will remain:

> Accumulating evidence suggests that *Lycopersicon* may not stand as a genus distinct from *Solanum.* If they are combined there is no chance that *Solanum* will be dumped for *Lycopersicon,* so even though the name *L. esculentum* has been conserved, if the tomato is treated as a *Solanum* its name will become *S. lycopersicum,* which will provoke howls of anguish from agriculturists and other non-taxonomists. We could conserve the name *S. esculentum,* which would go halfway toward solving the problem, but that would still leave our critics rabid. And there will be no escape, because that name-change will be essential to reflect accurately that opinion about the relationships of the tomato. Worse, because reasonable people may differ in this matter, some taxonomists will continue to use L. esculentum, some will use *S. lycopersicum,* and the matter could continue unresolved for many years. (96)

The issue of different interests between the person in the field or in the marketplace and the systematist who ultimately arrogates the right to name appears again in the case of the Douglas fir: "Proposals to establish a list of *nomina specifica conservanda* have been presented at several successive congresses. Such lists tend to be favoured especially by foresters and agronomists, who are not much concerned with formal taxonomy. I can understand the outrage of foresters at the change of the name of Douglas fir from *Pseudotsuga taxifolia* to *P. menziesii,* all because of some bibliographic digging by a man who knew little about the tree" (Cronquist 1991, 302). In the case of tomatoes, we have differences between the scientist who knew too much and the practical field person; here we have anger directed at the armchair scientist who knows too little. The language is extremely strong in both cases—words like "outrage" or "anguish" and reference to "rabid" critics: clearly the authority to name is an issue that touches some raw nerves among those who feel that they have no use for the names being vested upon them. All last century—since the Vienna Botanical Congress of 1905—some names have been preserved against the ravages of the taxonomist. But after all, is what else is a taxonomist to do: "after the relative stability of the Code in the 1950s and 1960s, there has been an increase in rather important or even radical changes to it. A younger generation of aspiring nomenclaturalists has grown up and tried to make its mark; and how well it has succeeded!" (Stace 1991, 75). McNeill (1991, 119) argues against "consensus taxonomy" since this would be, he says, the death of the profession of taxonomist. P. J. James (1991) asks the key question: "Should biological nomenclature, in the interests of user friendliness, be a chronicle of the evolution of biological thought, a slavish handmaiden of taxonomic theory, or should it concentrate on being a stable information retrieval system?" (63).

What is going on here in informational terms applies to a range of disciplines and can be expressed simply, even if over one hundred years of international meetings have not come close to settling the problem. To keep track of results in the sciences, you need to be sure of what you are dealing with—a rose should be a rose should be a rose, whether in seventeenth-century Leipzig or twentieth-century Pesotum. So the first principle is that you need as much name stability as possible. However, new understandings of plants can lead to rearrangements of taxonomy: two genera seen as historically distinct—the Chinese cabbage and the European turnip, for example—might now be seen as one (Chauvet 1991, 36). In this latter case, both vegetables are of considerable economic importance, so it will be very difficult to change the name—further, it will require some painstaking indexing work to help you track the losing genus through the literature (it is akin here to the problem in history of following women through name changes due to marriage). This merging/splitting work is done at every level—from the species up to the kingdom (at which level, the disputes between the Zoological and Botanical Codes about who gets to name, say, protists, are called ambiregnal problems). However, if you do not change the names, then the naming system loses its connection to theory and becomes more of an arbitrary mapping of the world, which militates against the point of producing names in the first place. To make matters worse, this area is far more intense in its demands than some others. In medicine, where exactly the same theoretical issues arise with the International Classification of Diseases—actually a statistical classification and a nomenclature (Bowker and Star 1999)—we are dealing with far fewer entities than the number of plant and animal species. In general, there are machineries of difference that engage whenever a standard classification is produced, ensuring that local communities of practice will move away from a given standard (or, recursively, metastandard) over time.

The Time of the Name: Priority

The botanical community has developed over time a set of strictly defined bureaucratic procedures for dealing with naming. A basic principle is that of priority—often a bugbear in scientific communities, but here a particularly difficult issue. The problem is that in a widely distributed literature—across all the continents and a number of disciplines, each with their own sets of journals—you need a rule to decide which name you are going to standardize. Priority has its own metapriority as the solution of choice to such problems: however this can lead to standardizing a name that is in minority use and losing an almost universal name.

The principle of priority is rooted in the work of Linnaeus, who invented the binomial naming system, which provides a consistent mode of naming all plants. In general, it has operated on the principle that a plant name be recognized if it

- has been given after 1753;
- is in accord with the Linnaean system (a thorny issue; since many non-Linnaeans in the early nineteenth century gave names which nevertheless can be interpreted as of binomial genus/species form);
- is not invalid according to the current rules of botanical nomenclature. (Briquet 1934)[6]

A plant's full name evokes a detailed history, providing after the binomial an abbreviated reference to the author of the work where the name first occurred and a reference to the site of the publication. For example, the Douglas fir (figure 4.4) was originally *Pinus taxifolia* (the latter meaning "having needles like a yew tree"). This was published in 1803, but a homonym was later discovered

6. Alex Panchen, *Classification, Evolution, and the Nature of Biology* (Cambridge and New York: Cambridge University Press, 1992). Panchen notes that the zoological code is phenomenologically complex in that the date that is set for the start of the priority of names is not an actually date but an arbitrarily fixed date when two works are "deemed" to have been published. He quotes the Code: "'The date 1 January 1758 is arbitrarily fixed in this Code as the date of the starting point of zoological nomenclature. Two works are deemed to have been published on that date:

1. Linnaeus's Systema Naturae, 10th Edition; and
2. Clerck's Aranei Svecici

Names in the latter have priority over names in the former'" (115).

Figure 4.4
Douglas fir (*Pseudotsuga menziesii*).

from 1796. A new name (*Abies taxifolia*) produced in 1805 ran into homonym problems too. Then things got complicated:

> By 1895 it was well known that Lambert's *Pinus taxifolia* was a later homonym. However, if one considered Poiret's *Abies taxifolia* to be a new name for *Pinus taxifolia*, as Sudworth suggested, then the epithet "taxifolia" becomes available for use in *Pseudotsuga*. Sudworth then proposed *Pseudotsuga taxifolia* (Poir.) Britt. ex Sudw. in 1897. Unfortunately, at this time it was not known that even Poiret's *Abies taxifolia* was a later homonym! The stage was then set for a series of esoteric arguments over the correct name for Douglas fir. During the period from 1897 until 1938, three different scientific names were used for one of the most important forest trees in all of North America: *Pseudotsuga taxifolia* (Lamb.) Britt. or more correctly *Pseudotsuga taxifolia* (Poir.) Britt. ex Sudw., *Pseudotsuga douglasii* (Sabine ex D. Don) Carr., and *Pseudotsuga mucronata* (Raf.) Sudw.
>
> In 1938, two English taxonomists skilled in botanical nomenclature, Thomas A. Sprague and Mary L. Green, concluded that the correct name was *Pseudotsuga taxifolia*, based not on the illegitimate *Pinus taxifolia* Lamb. (1803) but on *Abies taxifolia* Poit. (1805). (Reveal 2004)

After this, things got worse.

Thus this apparently simple set of criteria is notoriously hard to carry out in practice: one needs rules that are rigid enough to allow an unambiguous determination of difficult cases and yet flexible enough to accept publications which do not strictly follow the rules and yet have led to a universally accepted name. The first rule that was developed to standardize naming was the so-called Kew Rule, applied by botanists at Kew Gardens in London in the 1820s. According to this rule:

> Only epithets that are already associated with a generic name are considered from the point of view of priority when that genus is being revised—this is the so-called Kew Rule. Priority dates from the time that the specific epithet is first associated with the generic name. Hence epithets that may be older, but which have only been associated with species placed in other genera, can be ignored, and when genera are combined, well-established names are less likely to be changed. The major issue in the Kew Rule is how priority is interpreted when genera are combined, a minor, but associated issue is that of whose names are to be cited when a plant name is transferred. (Stevens 1991, 157)

The epithet here is the distinguishing name for a species, often describing one of its features: thus the epithet "esculentum" in *L. esculentum* (the old name for the tomato) means "edible." The major effect of the Kew Rule was to give botanists some flexibility in renaming: if a plant was given a name but under what was now believed to be the wrong genus, then that name did not have priority. Priority only accrued if genus and species remained constant.

In the 1860s there were two large-scale projects to standardize naming: George Bentham's *Genera plantarum* and Alphonse de Candolle's "Laws of Botanical Nomenclature" (Stevens 1991, 158). Bentham strongly defended the Kew Rule and argued in general that there should be room in the canon for names that were not strictly correct in Linnaean terms but that were nevertheless widely accepted (Stevens 1991, 162).

Candolle's work included lengthy discussions on the nature of priority, drawing particular attention to the tension between the name as history and the name as signifier. His laws formed the kernel of the rules of nomenclature operating maintained through a series of international biological congresses in this century and still in operation to this day. He drew attention to issues of just how much history could be wrapped up in a name. Thus he cited Kirschleger's discussion of *Mulgedium alpinum L. sp. 117 (sub: sonchus), Less. Syn 142*. The 'L.' here refers to Linnaeus, who was not the namer of the genus Mulgedium—in fact the epithet *alpinum* was Linnaean and the full name was due to Lessing—named after the first parenthesis. Candolle pointed out that this long and complex name would frequently be shortened in card catalogs or other lists so to *Mulgedium alpinum L.*, which would give a false history but would still be an effective, unambiguous name. He argued in two ways that the history wrapped into the name did not serve the purpose of providing glory, but was rather a simple convenience for arriving at an identifier:

> When one wants to pay homage to a botanist, one dedicates a genus to him. When you want to speak about his merits or demerits on the subject of a given species or genus, one adumbrates and discusses his opinions either in the text of a description or by some parenthesis in a synonymy—however the citing of a name or names in a plant name does not in itself express either merit or demerit. It is the statement of a fact—that is to say that such and such an author was the first to give such and such a name to a genus, or that he was the first to attach this species to that genre. (Candolle 1867, 47)

He argued that "the name is what counts most. . . . You might change whatever you want in your description of the genus *Xerotes* Br.; but one thing is fixed and certain and that is that Brown, in 1810, designated a genus with this name" (Candolle 1867, 53). He pointed out that the traveler who first picked the plant should perhaps be rewarded, but that priority in publication was the naming rule (ibid., 54). Candolle even looked forward to a future day when the current set of names might drop out; once science had succeeded in describing definitively what plants there were on the face of the earth, then the current "scaffolding," which contained many local exceptions to strict rules, might fall away. This however, was not for the immediate future (ibid., 7). Here he is echoing

Comte on scientific classifications, and indeed the French revolutionary calendar on the dating of events: when scientific precision is introduced, the history can drop away. One might draw a line somehere in the eighteenth century between those who sought a perfect language (unecmbered by indefinite articles and adjectives) in a golden age in the past and those since who have sought it in the proximal future. In either golden age, utterances say exactly what they mean and do not need qualifiers (context).

The flexibility advocated by Bentham and Candolle enraged Otto Kuntze. He produced in 1891 a list of 30,000 names that would have to be changed under a strict application of the laws of priority—and he wanted these changes to be effected (Briquet 1906, 5). Kuntze excoriated proponents of the Kew Rule and called Bentham, in particular, "a great sinner in nomenclature" (Stevens 1991, 162). Though many in principle admired his work, its root and branch changes were considered in general impractical. They led directly to the Berlin Rule, according to which generic names not in general use fifty years after their publication could be abandoned. In 1906 Kuntze stormed out of the International Botanical Congress in disgust, claiming that his protests were not being taken seriously enough (Wiesner 1906, 112).

Jean Briquet was a prime mover at the International Botanical Congress in Paris in 1900, and he headed the commission set up by that congress to determine a new code of botanical nomenclature. Priority remained central, but it was palliated by the "conservation" of certain names that lacked priority but had universal acceptance. Needham (1910) commented on the work of the commission:

> We have accepted the alteration of hundreds of well-known names that are root-names of many more genera within their respective groups: and such derived names, once of great assistance to the memory, have, so to speak, the props knocked from under them.
>
> Finally, and most lamentably of all, by our hasty and profitless abandonment of even the best-known family names we have broken with our best traditions . . .
>
> The pursuit of stability through rules of priority that has led to all this is surely one of the most singular of contemporary psychological phenomena. (296)

I will return to Needham's reference to memory later in this chapter. What is significant here is that priority is seen by Needham as an unnecessary principle indicative if anything of psychological disorder: "Why should it [the international commission] determine merely whether a certain forgotten name, abandoned by its author and never used, is really eligible for use under the rules of the code? It grieves me to see fifteen big brainy men, capable of doing something rational, put into a hole where they are expect to do only such little sinful things as this" (ibid.). The word "sinful" is evocative of the passion of the priority debate, echoing as it does Kuntze's charge against Bentham.

Throughout last century, the International Congresses ever further refined in parallel the application of the rules and the granting of exceptions to them. A typical entry in a relatively recent (Anon., 1965) proposal for the conservation of a name gives some ideas of the kind of work—at once botanical and bibliographical—that is involved in the maintenance or breach of priority. B. Verdcourt from Kew proposed conserving the generic name *Warburgia* Engl., 1895, against *Chibaca* Bertol. F., 1853 (*Cannellaceae*). He argued that *Chibica* was considered invalid soon after its publication, and that Bentham and Hooker had added it to their "genera ramanent indefinita et nomina delanda" ("genera remaining undefined and deleted names"). In an Italian journal in 1937, Chiovenda suggested that *Chibaca* was a member of the Canellaceae and was identical to *Warburgia breyeri* Pott. Chiovendi had not actually seen any *Warburgia* type specimens though—and so Verdcourt held fire. Then in 1964 he was passing through Bologna and stopped to search for Chiovenda's type specimen, but it could not be found. German troops lodging at a farmhouse near the herbarium had burned the collection. Verdcourt went back to the literature with his interest piqued and determined that *Chibica* was *Warburgia* but that the latter should be conserved, since the former name "is virtually unknown and has never been used in any flora or paper other than that by Chiovenda" (ibid., 27–28). Thus in order to effectively breach the principle of priority, Verdcourt had to prove that priority should apply according to the rules of nomenclature and then make the argument that in this case it would cause unnecessary problems to actually apply it. Equally, when the rule is applied, it is often necessary to do intense bibliographical work; Galtier (1986, 6) notes that it was necessary now to get down to the day and month of publication in the nineteenth century in many cases.

Priority, then, has been seem by some as everything from a pure naming convention (Candolle's position) to a matter of grave importance (Kuntze)—the lines are still drawn to this day. The list of botanical names willy-nilly serves as both an honor roll (and in so doing necessarily contains a highly formalized and abbreviated account of the history of each name) and as a set of arbitrary identifiers (since everything must be called something). There was a short-lived attempt to introduce the concept of "numericlature" (giving each taxon a universal number) rather than a name, but this did not gain many adherents (Little 1964): names are taken much more seriously by scientists and by the general public than are numbers or arbitrary identifiers, so that rather than solve the problem of naming it would merely have added another layer to its complexities (cf. Bowker and Star 1999, chap. 2, on alternative naming schemes for viruses).

The Space of the Name: Effective Publication

The issue with priority is who came first; the issue with publication is where did they come from? In the early nineteenth century, when there were fewer scientific journals, the most general problem was dealing with works not in English, French, or German. Over time, the number of journals has increased dramatically, and so the amount of bibliographical work that must be done to locate and propagate a name has risen in conjunction (Gunn et al. 1991, 279–280).

Candolle's first principle of botanical nomenclature, accepted by Briquet as part of the International Code and still in place, is that "Natural History can make no progress without a regular system of nomenclature, which is recognized and used by the great majority of naturalists in all countries" (Candolle 1867, 13; Briquet 1906). He discussed the problem of referencing publications within plant names. With that passion for systems that characterizes much work in this field, he discussed the problem of author abbreviations in botanical names and enumerated some 47 vowel and diphthong combinations that could hide in between the *h* and the *k* in *Hkr:* and then pointed out that the same 47 could hide between the *k* and the *r* leading to 2,209 possible names (Candolle 1867, 56). Just as he argued that the name of the plant was just a name and not an attribution of glory, so he argued that the publication of a name was just a publication and not something still owned by its author: "Can an author who regrets having published a name change it? Yes, but only in those cases where the names could be changed by any botanist. In effect the publication of a name is a *fact* that the author cannot revoke" (ibid., 57). Candolle did not, however, discuss just what a publication was.

By the time of the Paris Congress in 1905, the definition of "publication" had become an important issue—complicated of course by the fact that the further one went back in time the less well-defined was the field of scientific publication. (I note in passing that many of the rules adopted by botanists with respect to nomenclature can be read as an attempt to apply retrospectively whatever the current canons of scientific publication were to previous generations—which inevitably led to distortions of the historical material and so to a kind of active reading in science that would only be developed in literary criticism in the mid-twentieth century—a movement countered, for example, by Candolle's enunciation of the principle of "never making an author say what he has not said" (Stevens 1991, 159). Briquet's Commission proposed the following definition: "Publication is effected by the sale or public distribution of printed matter or indelible autographs. Communication of new names at a public meeting, or the placing of names in collection or gardens open to the public, do not constitute publication" (Briquet 1906, 53).

Equally thorny at that period was the question of whether "diagnoses" (formal descriptions) of plants had to remain in Latin. This problem was raised by the Spanish delegates, who wanted their language accepted alongside French, German, English, and Italian as legitimate languages for a diagnosis (ibid., 131). They lost the battle, on arguments such as that by Professor Maire:

> The principle of an obligatory Latin diagnosis . . . is the only means of conserving at present an international language, which language is an immense privilege for systematic botany. If we admit diagnoses in three modern languages, then everyone else will want to join in: after the Chinese there will be no reason to refuse the Papuans, the American Indians and all peoples who may one day accede to scientific life. Systematic botany would become a veritable Tower of Babel. (Brooks and Chipp 1931, 583)

There is an element of irony in the enforcement of Latin in the interests of internationalism—but it is unclear what alternatives there were. Many botanists were concerned about the discovery of plant descriptions in valid form in Russian, for example, supplanting (as it were) current names in Western Europe. This could be particularly difficult if different philosophies of naming were in operation—for example, it has been suggested that in Marxist Russia there were no infraspecific categories because these were not acceptable to dialectical materialism's insistence on the irreducibility of species (Heywood 1991, 54; cf. Graham 1972).

Over the course of this century, the issue of how many journals to look in has been problematic. At Kew Gardens currently, some seven hundred journals "are regularly scanned, as well as monographs, floras and other works in which new names might be found" (Lock 1991, 287). Each new discipline that has grown up has spawned new journals. Fossil species have long presented difficulties: for example, the *Journal of the Geological Society of London* is only rarely read by neobotanists (those concerned with current flora), yet paleobotanists have proposed new fossil genera in it (Boulter et al. 1991, 238). From January 1, 2000, it has been proposed that "the names of newly described botanical (including fungal) species will have to be registered in order to be validly published," and the Clearing House Mechanism is being adapted to coordinate this on an international basis (Heywood 1991, 57).

The issue with publication, then, has been how to be sufficiently universal so as to accept all scientific work done throughout the world, and yet sufficiently restrictive so as to make the information management problem tractable. Frequently, the latter criterion has been met by regularly underrepresenting work not in English or some other major European language and by ignoring work not in an ill-defined set of central journals.

The Sound of the Name: Euphony

Euphony has been a surprisingly resilient problem in the history of the naming of plants. Linnaeus has as one of his basic recommendations that plants, in order to be easy to remember—"all botanists must know and remember all the genera" (Cain 1958, 144)—should be easy to pronounce; they should be euphonious. George Bentham in 1838 observed that "names that seemed very difficult for an Englishman to pronounce might be easy for a Pole, Russian or German, and *vice versa*" (Stevens 1991, 160). But by the next century it was back on the agenda. In the 1905 Vienna Congress (one of the turning points in the history of botanical nomenclature), Linnaeus's recommendation is echoed in the following:

> V. Botanists who are publishing generic names show judgement and taste by attending to the following recommendations:
>
> . . .
>
> c) Not to dedicate genera to persons who are in all respects strangers to botany, or at least to natural science, nor to persons quite unknown.
>
> d) Not to take names from barbarous tongues, unless those names are frequently quoted in books of travel, and have an agreeable form that is readily adapted to the Latin tongue and to the tongues of civilized countries. (Briquet 1906, 39)

Each line requires a little elaboration. The "judgement and taste" phrasing is there to emphasize that this is a recommendation and not a requirement: the congress was attempting to deal with the problem of consistently naming all taxa over the whole world over all future time, and so wanted to keep requirements to a minimum. I have added (c), which is not a principle of euphony but does give an indication of the company that euphony kept: adjurations to civilized behavior in contemporary terms, and ways of excluding the outsider and the underdeveloped in more recent coinage. As late as 1971 a new botanical nomenclature (NBN) was proposed that would preserve euphony in similarly ethnocentric fashion. The NBN uses Esperanto, where "the words are pleasing to the ear, [and] there is enormous flexibility in word-formation, etc." (Smet 1991, 180).

Thus one person's euphony is another's cacophony; and yet as with priority and publication, it is a prima facie reasonable requirement. The problems have arisen—again as with the others—when you try to turn a set of precepts that have worked for a loosely defined club of largely Western European natural philosophers into a system that can work universally.

The Reach of the Name: Organizational Dimensions

The naming system in botany has served as a means of demarcating professional and research communities one from the other. Candolle (1867) made clear the distinction between botany and zoology: "[Linnaeus's system] has often been cited in philosophy courses. It has been considered superior to chemical nomenclature, because it lends itself better to changes necessitated by progress. Botanists professed a veritable cult for the system. They prided themselves on having better understood and developed it than the zoologists" (3). Indeed, his first principle of botanical nomenclature, taken up in the first international code and still the first principle to this day, was that "botanical nomenclature is independent of zoological nomenclature, in the sense that the name of a plant must not be rejected merely because it is identical with the name of an animal" (ibid., 12); and there are indeed cases of plants and animals with the same Latin names. This distinction has led to a series of border disputes concerning just what should or should not be included in the nomenclature (figures 4.5 and 4.6). One thorny issue has been ambiregnal species—species that might equally well be designated plant or animal. This has been an increasing problem with new phylogenetic work "increasing the number of major new inherently ambiregnal clusters of autotrophic (plant-like) and heterotrophic (animal-like) protists" (Patterson and Larsen 1991, 197–198). Most believe that generating a single code is just not going to happen, but that arbitrarily assigning protists to one code or the other is equally problematic (Patterson and Larsen 1991, 201).

A dispute erupted with bacteriologists at the fifth Botanical Congress. The bacteriologists, led by Buchanan, wanted an exemption from the need to use

Figure 4.5
Prunella modularis. Source: http://www.povodok.ru/pics/art34/art3392_0.jpg.

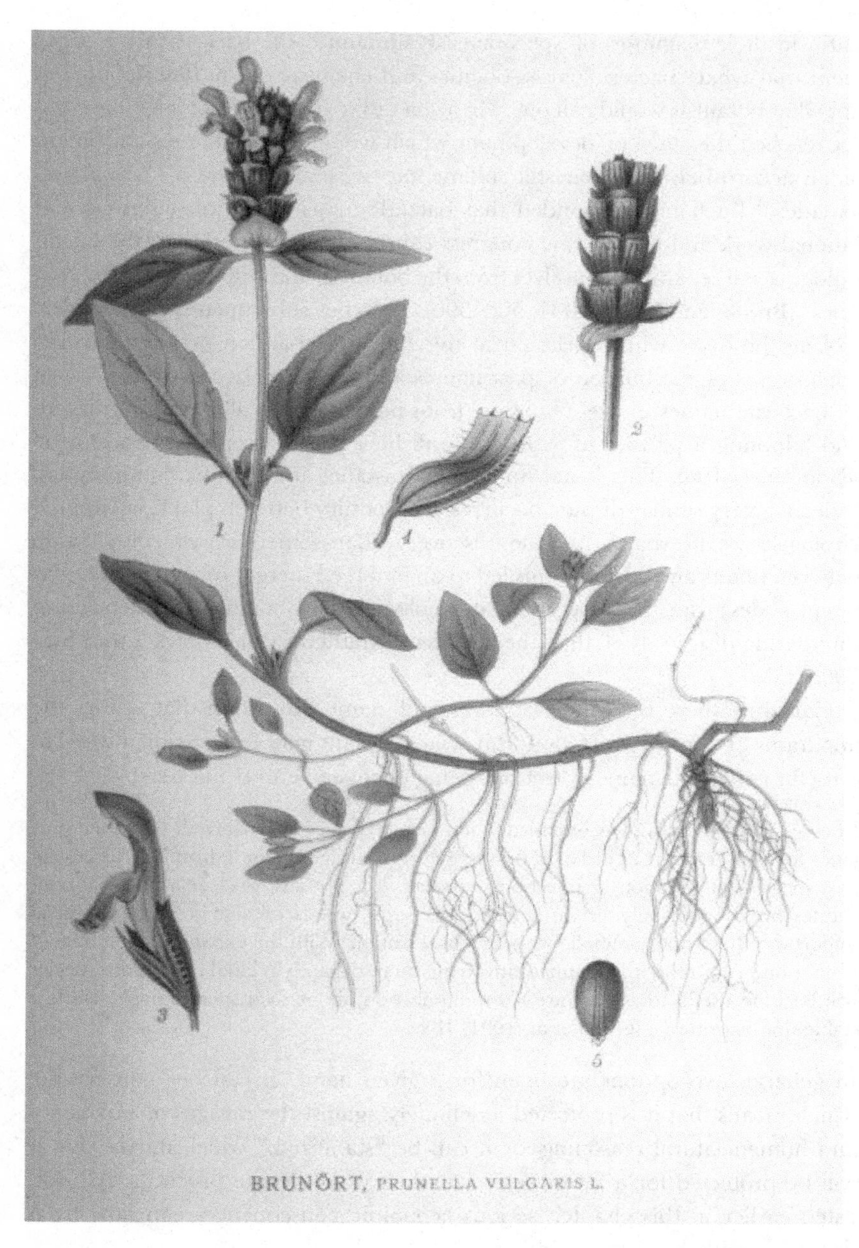

Figure 4.6
Prunella vulgaris. Source: http://www.planetbotanic.ca/images/selfheal.jpg.

Latin in their diagnoses of specimens. Haumann came back with the argument that what "bacteriologists, doctors and chemists call a 'description'" is not what botanists would call one. He argues that "the bacteriologists have not yet reached the stage of development which would permit the establishment of an accord between their still rudimentary systematics and a rational systematics." Buchanan responded that bacteriologists were doing serious and rational work and that "if the congress cannot accept my motion, the bacteriologists will separate themselves from the botanists and will develop their own rules" (Brooks and Chipp 1931, 588–590). This they subsequently did—underscoring the move with a decision in the 1960s to abandon priority and free "themselves of the burden of past names and literature by adopting a list of all bacterial names in use, removing from nomenclature all names not listed, and adopting a process of registering all new names proposed henceforth" (Ride 1991, 106). This is not an isolated instance in scientific communities. Indeed, a very similar dispute occurred this century between plant and animal virologists—exacerbated by the discovery that some viruses could jump between plants and animals: this led to an enforced merger of two fiercely different codes, with the proponents of zoological and botanical nomenclature thundering dismissals of the others' system (Mathews 1983; Bowker and Star 1999).

Currently there is a whole apparatus of name protection that echoes the apparatus of species protection; thus when a plant goes from being a weed to a useful variety, its name goes from being changeable to being fixed:

> In cases where names of economically important species are involved, the Code provides for conservation in order to preserve current usage. Judging from the success of past species conservation or rejection proposals, it is not always clear as to what constitutes an economically important species. Some weeds of significant agricultural importance have not qualified for name conservation. With the expanding potential of gene transfer for crop improvement involving more distantly related taxa, more species will become useful to agriculture. Communication about such species depends on a stable nomenclature (Gunn et al. 1991, 18).

In general, two options are open for a given name: it can be "conserved," which means that it is protected indefinitely against the ravages of taxonomy and nomenclatural reasoning, or it can be "stabilized," which means that it will be protected for a given period of time while debate proceeds. As indicated earlier in this chapter, serious economic consequences can flow from decisions made by taxonomists.

The problem of cultivars (cultivated varieties) and their naming has been a constant one: in general, it has been remarked that botanical "snobbery" has meant the overlooking of "substantial horticultural works of the late nineteenth and early twentieth century" (Mabberley 1991, 125) as published sources for priority purposes.

Overall, the situation today is in some ways much the same as it has been over the past two hundred years. An author recently noted: "For any given conserved tropical wildland we are confronted with a problem roughly analogous to receiving an enormous library with no call numbers, no card catalogue, and no librarians—and the library being in a society that is only minimally literate and not even certain that reading has much to offer. The library is hardly more than highly flammable kindling in such a scenario" (Klemm 1990, 23). The image of the burning library cropped up again in 2002 in the Taxacom Discussion Listserv. Here is an extended quote from Russell Seymour that is redolent of Lyell's concern with archives commissioners:

> Consider that a proportion, perhaps the greater proportion, of the books rescued from the conflagration in the library and stored in the vault are written in a 'dead' language; a language that has no living readers or speakers. (It could also be that the language is still out there somewhere, but no-one is able to link the book with the language.) We may be able to recognise the letters, and that they are constructed into words and sentences, but without the context of the grammatical rules of the language our ability to interpret the information contained in these books is severely curtailed. Of course these books should be cared for and held into the future. Perhaps one day a 'code-breaking' linguistics expert will recognise a pattern to the letters and will pull out more information, but without the knowledge of vernacular language and the subtle nuances of every day use information content will be lost, or, perhaps, completely misinterpreted . . . So the analogy is that the books are species. The language they are written in is their 'context' representing the species' habitat and all of the ecological (biotic and abiotic) interactions that involves. The death of this language does not necessarily imply the loss of the habitat, but represents the removal of the book (a.k.a. species) from its context (a.k.a. habitat); effetively the extinction of the species. The vault is still the museum collections. The 'linguistics expert' is your friendly neighbourhood gene-jockey with his DNA sequencing widget in his back pocket. Letters, words and sentences are nucleotides, genes and gene complexes. . . . The bottom line is that once that book has been separated from its language it is a hell of a job to understand just what it is all about. . . . So what should our 100 volunteers do? Should they be water carriers attempting to douse the flames or should we spread the resources and take 50 of them in to rescue the books? Can we have our cake and eat it too? Consider that the 100 bucket carriers, once they have thrown their water into the inferno [. . .] now have empty buckets, and they are stood next to the very shelves on which our books are sat. Why not get them to grab a few books, put them in there bucket and hand them over to be taken to the vault? The bucket-wielders may not know what they are retrieving, but that is the job for the guys in the vault. (Taxacom Discussion Listserv, 8/22/02)

The burning library metaphor sparked many messages in the list; it speaks to the convergence of conservation in nature and in second nature (typified here by the museum and the library).

In the context of the conflagration, renaming can be highly problematic. When subspecies or varieties are elevated to full species rank: "This may have very unfortunate consequences from a legal point of view when the species to which the subspecies or variety belonged before the nomenclatural change is listed as a protected species, the result of the split is that the new species looses its protected status unless the legislation is amended to add it to the list" (Klemm 1990, 33; cf. Lucas 1993, 11, on the Flora Europaea). There is a move to create an internationally sanctioned register of plant names, but that move itself has run up against First World/Third World issues, since key gatekeeping control would be retained by the industrialized nations. In September 1998, a group of Argentinian botanists wrote to the International Association of Plant Taxonomy prior to the July 1999 meetings in St. Louis to protest the attempt:

> We want to express:
>
> Our strong support to the present **International Code of Botanical Nomenclature**.
>
> Our opposition to the **Registration of Plant Names**, which will introduce more difficulties than advantages, especially for countries outside Europe and the United States. Also, it should be mentioned that this project will add bureaucracy to the existing requirements for publication of benefits without significant gains. In addition, we believe the principle of priority should remain in the way it is right now.
>
> Finally, our support of an open and democratic election of members of the board of IAPT, avoiding discretional preparation of the list of the proposed officers by a few associates. (http://mason.gmu.edu/~ckelloff/vfunk/latamsig.html; accessed June 2, 2001)

The move to register all names, to agree on model data structures and formats for biological databases in order to facilitate biodiversity management (Heywood 1997, 12) is just as urgent and just as overly optimistic as the calls of Candolle for a rational system of nomenclature. Currently, there is a continuation of the proliferation of names for plants in different organizational contexts: "The net result of instability in names has been an increasing tendency for consumer organizations to issue their own standard lists of names fixing species names, at least for set periods of time. Examples are the CITES convention concerning trade in furs and plants . . . , the International Seed Testing Association, and FAO/WHO lists of organisms of quarantine importance" (Hawksworth and Mibey 1997, 19).

In short, the production of consistent names for all plants is a very rich organizational and intellectual process. It has over time involved

- setting up rules for the reading of documents—and indeed producing an understanding of just what kind of activity reading is;
- deciding just what kind of a thing a publication is;
- endeavoring to find a name with a pleasing, memorable sound (and to deal with cross-cultural issues in deciding euphony);
- negotiating with other scientific groups (zoologists, bacteriologists) and with commercial and regulatory bodies (horticulturists, nurserymen, government agencies);
- dealing with differences between developing and underdeveloped countries about control of naming conventions.

This set of issues is matched by other bodies (e.g., epidemiologists) who try to maintain data sets for extended historical periods and geographical sweep. They are issues that have generally been ignored each generation as the new set of information technologies is brought into play along with its particular dream of a common language (Rich 1978). Naming is a difficult thing to do. It is a site of important decisions for the organization of knowledge and for the organization of scientific work; and it is an activity with political and economic consequences. These dimensions should be fully factored in to the development of new information systems to deal with the burgeoning huge data sets that are a necessary adjunct to biodiversity management.

What's in a Name? We have in this section on naming looked in turn at things that are hard to classify, things that do not get classified, and things that are classified in multiple ways. Within each category, we have seen organizational and political concerns rubbing up against scientific and philosophical questions. The activity of naming is fundamental to shaping our databases of biodiversity information, but it cannot be carried out effectively without the construction and maintenance of a series of organizational procedures. The names we choose for our biodiverse world tell organizational stories and contain within them the seeds of historical narratives.

The Problem of Context

Let us assume for the moment a perfect world—always a pleasant interlude in a book replete with bureaucratic entanglements, wild politics, and theoretical conundra written against a backdrop of a mass extinction crisis. In this perfect world, we have solved the naming problem: names mean just what we want them to mean and nothing else, and they allow us to tell stories that are just so.

Lest the reader relax too much, be assured that there are clouds looming on the horizon. For even if we can name everything consistently, there are the problems of how to deal with old data and how to ensure that one's data doesn't rot away in some information silo to paraphrase Al Gore's memorable phrase (President's Committee of Advisors on Science and Technology 1998) for want of providing enough context. The problem with biodiversity data—as with much environmental data—is that the standard scientific model of doing a study doesn't work well enough. In the standard model, one collects data, publishes a paper or papers, and then gradually loses the original data set. A current locally generated database, for example, might stay on one's hard drive for a while then make it to a zip disk, then when zip technology is superseded, it will probably become for all intents and purposes unreadable until one changes jobs or retires and throws away the disk. There are a thousand variations of this story being repeated worldwide—more generally along the trajectory of notebooks to shelves to boxes to dumpsters.

When it could be argued that the role of scientific theory as produced in journals was precisely to order information—to act as a form of memory bank—as, for example, by Poincaré (1905), this loss of the original data was not too much of a problem. The data was rolled into a theory that not only remembered all its own data (in the sense of accounting for it and rendering it freely reproducible) but potentially remembered data that had not yet been collected (for the concept of potential memory, see Bowker 1997). By this reading, what theory did was produce readings of the world that were ultimately data independent, if one wanted to descend into data at any point all one had to do was design an experiment to test the theory and the results would follow.

The observation that we are dealing both by necessity and by choice with data stores as fundamental to biodiversity science does not of course mean that theory drops out of the equation. Indeed, as shown in the section on classifying, classification systems are theoretically shaped in their inception (by reflecting a set of theoretical beliefs) and shape theory in their deployment in databases (by making it easy to follow certain theoretical paths and hard to impossible to follow others). What we are frequently left with is a set of legacy data housed in legacy systems—botanical information stored in a hierarchical database and using an outmoded classification scheme for example. Clearly, it is often just not worth the effort of first massaging the data into a more contemporary classification and then migrating it across into a new format. This is particularly true since what goes for organisms goes for data: the vast swathes of data held in mundane format on old systems receive neither the theoretical attention nor the funding of esoteric data sets held in expensive new equipment. Thus, for example, Judy Weedman (1998) has written about the problem of getting computer scientists interested in solving the kind of problems that environmentalists handle: they are just not seen as theoretically interesting.

In this section, I lay out some basic ways of understanding legacy data. I will treat the issues that arise by looking in turn at the Scylla and the Charybdis of long-term data management: not keeping enough of the past and keeping too much of the past. One continually chases the other's tail. Both resolve practically into the single observation that if a legacy data store does not retain its own context as a formally separable set of entries, then it is useless. More deeply, they resolve into theoretical questions of just what do we know about the past, and how do we know it. We do not want a world with perfect memory:

• *Perfect memory has a high overhead*: It would be nice if we could preserve all the external murals in Italy, but this militates against our action in the present (we can't paint our current houses).

• *Perfect memory is not what it seems*: The Ise Shrine in Japan has been torn down and rebuilt every twenty years since AD 652 using the same tools and skill set; it is recognized as the oldest temple in the country. What is being preserved here is not the *ding an sich* (which creates a legacy of preservation techniques) but the mode of building (which creates a legacy of organizational forms). The overhead problem of course recurs at this level.

• *Perfect memory does not matter if no one is listening to your stories*: The "archival literature" in science is written as if someone someday will have time to go back and read all this welter of material and make sense of it—assigning priority, determining value, and so forth. This is the secular version of the Last Judgment—and is equally dependent on an Entity capable of massive data storage and analysis. There is no evidence that this Entity is in the process of formation.

Not Keeping Enough of the Past

Bowser (1986) has a fine description of the problems of dealing with old data—from what is now an NSF-funded Long Term Ecological Research (LTER) center in Wisconsin. In 1898 E. A. Birge, C. Juday, and their associates established a limnology group at the University of Wisconsin at Madison. Subsequently, research at Trout Lake Station was initiated in 1925. This is a region with one of the densest concentrations of lakes in the world, and with a number of accessible bogs and wetlands. Bowser relates that there were seven decades of data available to modern researchers—though most was concentrated in the period 1925–1941. He describes four major problems with the older data. First was the issue of sampling the lake water. Although it was recognized as early as 1924 that there were seasonal variations in acidity and other key measurements, the lake waters were only sampled once per year, in the summer. Thus, in order to make meaningful comparisons, modern studies had to use only data from July through September. Second, there were terminological changes: what had been a distinction between "bound" and "free" CO_2 became a distinction between alkalinity and acidity. This was uncovered through a literature search of the old data. Third, a more difficult problem was that the pH measures would be different depending on whether they were taken in the laboratory on return from the field trip or in the field—loss of CO_2 in samples over a few hours changes the measurements. This information was nowhere mentioned in the published reports, but fortunately Bowser and colleagues were able to locate a retired limnologist who remembered the procedure. Fourth, they found that there was a shift in the kind of measuring techniques used for alkalinity—from methyl orange to electrometric methods—which caused a break in values that could be compensated for once the change was recognized.

These four kinds of problem are archetypal. At the time of data collection, one might

- change measurement techniques without proper records (new alkilinity measures);
- use current terms localized in place or time to describe the data ("bound" and "free");
- not record key data that is seen as part of normal behavior in a given community of practice (measurement time); and
- not realize that additional data is needed (measurement date in our case).

The first of these is trivial on the surface, though dealing with it is extremely difficult. It represents part of the normal work of producing metadata standards: one tries to script standards in a way such that changes are automatically recorded. There is the standard practical problem here (cf. Bowker and Star 1999; Fagot-Largeault 1989, for the medical analogue) that the person filling in the forms is keen to be doing things other than ensuring the perfection of the record, since he or she knows about the new techniques and will not be led into making any mistakes with the data set and has better things to do than through an altruistic gesture make things easier for notional future users.[7] The second and third problems are deeper and more difficult. They

7. Mike Twidale (personal communication) is developing the concept of designing for altruism to respond to this problem.

require an act of imagination on the part of the record keeper to place him- or herself in the position of any possible future reader. They might assume that the average user of the data will know such and such a term through their formal education or through apprenticeship—but as we will see, with biodiversity and other forms of interdisciplinary data, much less can be assumed than one might suppose. In essence, the record keeper is being asked to abstract the record set out of the historical flow of time—to provide enough information so that a limnologist from Mars (who presumably has been out of work for several million years) can come along and, from the data set and a sufficient command of English, interpret the data. Measurements taken as recently as in the nineteenth century using nationally accepted standards cause problems today, as a query to the Taxacom listserv indicates:

Are conversion tables from fathoms to meters (or linear nonmetric to metric) for different countries available anywhere on the WWW? For example, a Danish fathom is 1.883 m whereas the Imperial (British or American) fathom is only 1.829 m, but if there are web sites with this kind of information, I am asking the wrong questions as I have not been able to find them. Are there similar differences for other European countries? For example, what system does France follow? Germany?—Would this be the Bavarian system? At the smaller end of the scale, a Danish inch is (still) 26.112 mm and a Danish line (1/12 of a Danish inch) is 2.176 mm, whereas an Imperial inch is 25.4 mm and the line is 2.116 mm. These nonmetric measurements, especially fathoms ("orgyiar" in the Latin of Malmgren, 1867) and lines, occur in many older descriptions and station lists from the 1800s. (Taxacom Discussion Listserv, 3/22/99)

(One could say that this wasn't rocket science; however, even rocket science has had the problem of the confusion of imperial and metric measures in the form of the Mars Polar Lander crashing into the planet). Of course, complete transparency of old data is not possible. Indeed, what is being demanded of the dataset is precisely something which over twenty years of science studies have shown cannot be asked of the scientific paper—to stand outside of time. The fourth problem is positively brutal—the measurements that are made now are necessarily constrained by current theory, and there is no way of making allowances for future possible theories.

Bowser (1986) writes that overall "the availability of 'data' has generally not been the problem with historic data, it is the unequivocal documentation of techniques that has concerned us most" (174). What is needed is a record of processes as well as a record of facts. However, processes and facts cannot be in principle be disentangled, so we are never going to have a perfect data set wrapped in complete metadata. Moreover, the processes that we need to record in order to ensure the viability of data in the long run do not constitute an easily enumerable set: they include information about how classification systems are arrived at; what the local coding culture is; what techniques were used, and so forth—social, organizational, and technical processes all make a difference.

Keeping Too Much of the Past

Any data coding scheme contains traces of its own past. This is frequently buried deeply enough that it might not be apparent to a contemporary user—particularly from a cognate discipline and one untrained in the vagaries of that scheme—that an archeological effort is needed in order to uncover deep-rooted biases. A classic example here is Walter's study of the European bias in angiosperm classification. Walters (1986) demonstrated that Linnaeus used the folk classifications available to him at the time, and that these classifications in turn favored describing more genuses in a family for economically important plants such as Umbilliferae (the carrot family) than for chickweed. Inversely, there are many more species per genus in the economically important grass genera (Graminae) than in sedge (Cyperaceae). He notes that a New Zealander starting ex nihilo would order the same set of species quite differently. Bowker and Star (1999) have noted this European bias in the International Classification of Diseases (ICD)—a bias based on the Parisian origins of the current classification, and that through the history of the deployment of the ICD has led to a number of complaints from tropical countries, feeling that their disease entities were underrepresented. Walters (1986, 538) goes on to criticize evolutionary hypotheses that have been made based on number of species in a genus and number of genera (maintaining that the oldest genera having the most divergence) as being an artifact of the original folk classifications. This example of traffic between data structure and narrative is not a pathological case to be rooted out. To the contrary, it is very much business as usual in our classification schemes: one of the true tasks for metadata development is how to represent such archeological richness in our data coding.

Models using any given data set frequently contain hidden traces of their own past. This issue has been explored by Paul Edwards (1999) in a study of circulation models used in climate studies: he has shown inheritance from the 1960s to the present of tenuous assumptions in some models. Models developed today can suffer from the same kind of "presentism" that plagues much historical writing. Thus Scott Wing (1997) describes the "equable climate paradox." This paradox affects general circulation models, which use starting conditions based on current theories about ancient coastline positions and ice cover, together with the same equations that are used to predict current weather patterns. The postulated weather patterns are then averaged to yield a projected paleoclimate. The problem is that there is evidence that, say, using proxy measures such as palm tree distribution, inner continents are warmer than they "should" be. Indeed, Wing says, there is a "strong tendency for climate models to yield results that are more like the present than they should be" (ibid., 170). Thus we keep too much of the immediate past in representing the distant past.

Such data or modeling problems are frequently well known within any one given discipline: there has generally been some kind of apprenticeship process whereby the new botanist learns the history of Linnaean classifications or the climate modeler learns about the equable climate paradox. However, things become more complex in a field like biodiversity science, where numerous others might be using one's data or drawing conclusions from one's models. Any given complex model of a given biogeographic region at some previous geological epoch might be drawing on a large number of models and datasets, each with their own relatively recondite legacies.

Consider, for example, the various bases that there are for providing temporal scales for the events on the earth. A number of different "clocks" have been suggested over the past two hundred years. One person—for example, Bishop Usher—might count the number of generations reckoned in the Bible and multiply them by average generation span to date time since the creation of the earth. Charles Lyell might reckon events from sedimentary evidence—using this to put lower limits on the age of the earth that far exceeded the length of time that the generational clock indicated. Then Lord Kelvin might use the pearl of all the sciences—physics—to show that on the assumption of a molten earth at origin and the current indirectly measurable inner heat of the earth, one only had about 40,000 years to date events in—not the 400,000 or 400 million indicated by geology (Burchfield 1990). Which brings us into this past century, when there have been a series of other dating conflicts. One that has caused a lot of controversy in the past ten years has been using the rate of mutation of genomes to give a "clock" for species development. Following the theory that mitochondrial DNA was passed down only from the mother (recently shown to be only approximately correct, as many such dogmas are) and that it mutated at a constant rate (recently shown to be probably incorrect; and certainly requiring ranking of different parts of the mitochondrion as better or worse timekeepers), one could argue for a "mitochondrial Eve" for various species—producing a clock that throws key evolutionary events much further back in time than would be indicated by paleontological evidence in the form of fossil remains (Strauss 1999).

Two observations emerge from the difficulties of reconciling clocks over different scientific disciplines over time. The first is that there is a strong power component in the debates—she who controls the clocks controls knowledge. Lyellian geology was developed in opposition to literal creationist narratives, so the struggle for the clock was simultaneously an assertion of the rights of the scientist to speak to questions of the history of the earth. Kelvin's attack on the Lyellian clock was equally an assertion of the power of physics as a deductive science over geology (Burchfield 1990). More recent debates have been about whether to prefer molecular biological data over field data (e.g., Ernst Mayr (1988, 9) decries the "unpleasantly arrogant manner" of some molecular biologists talking to morphologists.)

The second observation is that if you want to create a biodiversity database using a commonly accepted standard timeline, then you are going to have serious problems. In some fields there are standards-setting bodies at work that relatively regularly produce consensus clocks. Within geology, one might decide to use the Geological Timescale produced by the Geological Society of America in 1989 or the timescale produced by Gradstein and Ogg—or a mixture, as in a UNESCO project where contributors of paleotectonic maps were directed to use a timescale as follows: "early Permian to Holocene ages are from Gradstein and Ogg, 1996, A Phanerozoic Time Scale: Episodes, 19:3–5. Late Carboniferous ages are from Harland et al., 1990, A geologic time scale: Cambridge University Press" (http://www.geomin.unibo.it/orgv/igcp/maps.htm) (figures 4.7 and 4.8). The reason for choosing a mixture is that one timescale might better reflect a given set of measurements than another (e.g., keeping an extinction event within a certain epoch rather than having it straddle two). This is not just a problem for reconciling clocks; it occurs at each level of data integration across disciplines and across agencies. In Utah GAP analysis described earlier, a veritable smorgasbord of naming standards was available; the ultimate decision was taken based on agency affiliation: "The exact number of species by taxonomic group varied among different agencies having management responsibilities in Utah. Given that Gap Analysis is a state-based information system, we selected the species list accepted by the State of Utah Division of Wildlife Resource" (Edwards et al. 1995, 4–3). Similarly, Scott Peterson discusses the wide range of federal U.S. agencies that produce taxonomic lists and notes that the Forestry Service, for one, has difficulties "because of the proliferation of regional lists of botanical nomenclature and symbols that aren't easily translatable across regions and to other agencies" (Peterson 1993). Chrisman and Harvey (1998) report that there was only a 3.5 percent overlap for maps of wetlands in Wicomico County surveyed by four different agencies.

With much biodiversity data, there is no point at which you can say that such and such is a bedrock standard: it's triangulation all the way down. A

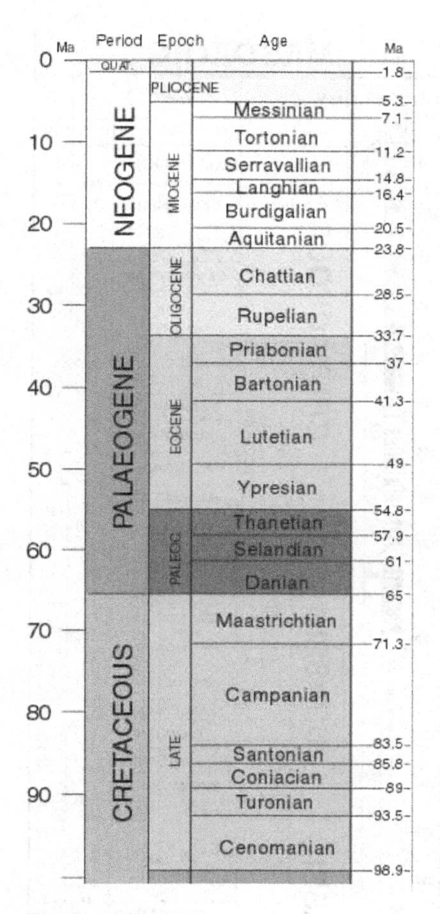

Figure 4.7
Gradstein and Ogg timescale, selected longs. *Source:* http://www-sst.unil.ch/igcp_369/Time-geology/timescale.htm.

given geological timescale might make sense because of a certain reading of the fossil record, itself coordinating among various kinds of surrogate measures, taphonomic theories, and current understandings of paleoclimates. In an article "Absolute Ages Aren't Exactly," Renne, Karner, and Ludwig (1998) point out that there are a number of decay constants in play for leading indicators of absolute age. Different disciplines produce their own incompatible absolute scales:

The decay constants used in the nuclear physics and chemistry literature are based on counting experiments of α, β, or χ radiation activity, whereas the values used by

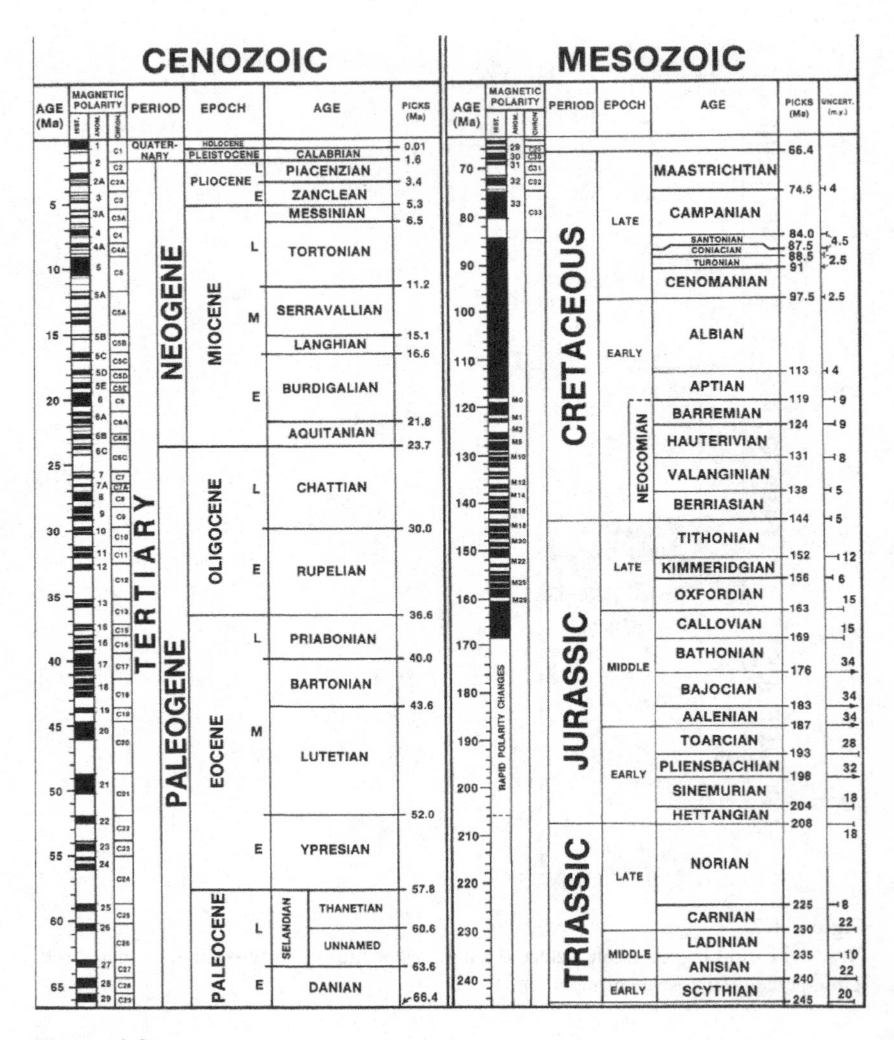

Figure 4.8
The decade of North American geology, geological time scale, selected longs. *Source:* http://www-sst.unil.ch/igcp_369/Time-Geology/timescale.htm.

> geochronologists also include, in some cases, the results of geologic intercalibrations or laboratory accumulation experiments. For ^{40}K, the two communities values for the total decay constant that differ by 2.1% simply because of different choices in filtering the same set of activity data. For ^{87}Rb, the value used by geochronologists, based on laboratory in-growth experiments, differs by nearly 3% from the value used in the nuclear physics and chemistry literature, a difference well beyond stated errors of the two values. (ibid., 1840)

They note that some key standards, such as of $^{40}Ar/^{39}Ar$ decay, have never received international review, and they comment that where there have been reviews of other standards, there has been poor interdisciplinary communication.

One can picture an ideal data storage system that would be aware of all the standards that had gone into the naming of its categories and the partitioning of its intervals. In this system, when the American Geological Society changed its timescale, paleoecological and paleoclimatological databases could be easily adjusted to fit the new standard. What's wrong with this picture? First of all, standards get deeply entrenched into infrastructures, so that as with the climate models referred to earlier, it can be very unclear to current users that such and such a standard has been implemented. Second, even if the paleoecologist hears about, say, the new timescale from the mitochondrial clock, he might not be at all willing to change his database to fit in with the new absolute timescale because such a change would throw off a series of other correlations (which may or may not be based in turn on their own sets of entrenched data). Finally, it is not only impossible to separate data structures from the internal theory of the discipline(s) whose data is being stored, it would also not be a good idea. For as we have seen with the case of the warring chronologies, there is a lot of good content and at the same time a lot of good political work being done at the level of deciding database issues.

The caution here is that with the process of the mutual imbrication of data standards, each of which covers its own context (its own past) in archeological layers or through radical renaming, there are certain secular processes of information convergence that entail that some kinds of narrative get progressively excluded from the picture in the same process that creates ever more robust (though always necessarily incomplete) mutual bootstrapping of the rest. If we want to create data structures that are more open—though none will ever be completely so—then we need to do two things. First is a very practical step: we need to retain the context of development of a given database in reasonable detail; the political and social and scientific contexts of a set of names and data structures are all of interest. I emphasize reasonable detail here: a perfect archival system is a chimera. Second, the goal of metadata standards should not be to produce a convergent unity. We need to open a discourse—where there is no effective discourse now—about the varying temporalities, spatialities and materialities that we might represent in our databases, with a view to designing for maximum flexibility and allowing as much as possible for an emergent polyphony and polychrony. Raw data is both an oxymoron and a bad idea; to the contrary, data should be cooked with care.

Information Integration

Biodiversity and Ecology vs. Biodiversity and Systematics There has been a move in a number of sciences over the past few hundred years to analyze basic scientific units in terms of information storage and transmission (Bowker 1994): be this genes, quarks, or species. This move is a point of articulation for two divergent memory practices in biodiversity research. By the simplest definition, biodiversity is just about the number of species that there are—one assigns a unique identifier to each species and then counts the number of species in a given unit area to get a biodiversity quotient. Many practitioners of biodiversity science have argued that such a measure does not give an indication of "true" biodiversity. They point, for example, to the case where one has a dozen species of rats and one of pandas (Vane-Wright et al. 1991)—note the charismatic megafauna being pitched against the ever unpopular, though highly nutritious, rat. The argument is made that in terms of information held in gene stock, the panda well outweighs the several rat species, which contain a set of overlapping genes. Formally, this means taking diversity as "a measure of information in a hierarchical classification" (ibid., 237)—with the implication that one wants to save ancestor species and species with few close relatives for preference over more recently evolved species with many siblings. As Barrowclough (1992) indicates in a detailed example, this line of argument can lead to a series of nonobvious choices; in an example, he remarks: "some of the taxa in the coastal forest of southeastern Brazil have a sister group relationship to other species throughout the Amazon basin and hence are in some sense equivalent to that entire avifauna" (137).

There is an interesting convergence between the world and its information at this point. Both in terms of databases on computers and the world as database, scientists are seeking the minimum data set that is needed in order to preserve biodiversity—whether that data set be the genes held in organisms or the bits held in computers. In both cases, this is not seen as the ideal outcome; to the contrary, it is seen as a practical choice—given that we are destroying biodiversity, and given that we do not have enough systematists. The estimated completion of the Flora Neotropica, begun in 1968, is the year 2397; this is not commensurate with the rate at which the environment is changing (Heywood et al. 1995). Stork (1997) notes that we seem to be in the middle of an extinction crisis and yet only about 1,000 species "are recorded as having become extinct in recent years (since 1600)" (45). He contends that it is hard to estimate loss, since in any case most species are only known through a single example—their holotype. Indeed, to continue the convergence theme, we could note that many of these holotypes are only known through books or other publications, since the original specimen either was not collected or has been lost: these are called lectoholotypes. In general, it can be argued that such a convergence leads, through the deployment of a common set of metaphors and methods, to a close resonance between the world and its information: the two are conjured into the same form. For our immediate purposes, the assertion that biodiversity should be seen as an information issue entails that strategies for both data collection and habitat management are intimately wrapped up in ontological questions about what kind of a thing biodiversity is. It is ironic that while computer memory is happily outpacing Moore's law, the information-carrying capacity of the earth (in the form of genetic information in species) is diminishing at an inverse exponential rate.

On the one hand, then, we have a set of information collection strategies twinned with biodiversity protection strategies based on the view that species are information units in a genealogical hierarchy. Eldredge (1992) argues that this is in contrast to an *ecological* perspective, which he ties back to an economic hierarchy, thus going back to the common roots of ecology and economics in the Greek word for household, *oikos* (Williams 1983, 110). The distinction works as follows. Ecological diversity reflects "the number of different sorts of organisms present in a local ecosystem" (Eldredge 1992, 2). Now a species does not operate as an economic unit—indeed, a given species is generally a member of a number of ecosystems. Ecological diversity (the number of species in a given community) is orthogonal to biodiversity. Eldredge argues that species are part of the Linnaean hierarchy, whereas local ecosystems are part of interacting economic systems. The former are extended in time and genealogical—they "act as reservoirs of genetic information" (ibid., 5)—whereas the latter are extended in space, their temporal hallmark being "moment-by-moment interactions." At base, then, Eldredge is arguing that we are dealing with two entirely different ways of being in the world: demes (subgroups of a particular species living together) obtain spatially and are part of ecological diversity whereas species obtain temporally and are part of biodiversity. Similarly, within biogeography, Grehan makes the assertion—challenged by Cox (1998, 821) that there are "dichotomies of ecology v. history, and dispersal v. vicariance" (Grehan 1994, 461).

By this persuasive pitching of ecology against history, the complexity of integrating biodiversity data across multiple disciplines is increased since it leads to a rift in the data collection efforts. Thus Wheeler and Cracraft inveigh against the concept of the All Taxon Biodiversity Inventory (ATBI), which has been an influential model in recent years. In an ATBI, an area is marked off and a group of taxonomists and parataxonomists work on inventorying all the species in that area. The authors argue: "While periodic collecting at known sites is a prerequisite for documentation of the status and trends of biodiversity, the very notion of long-term study at a few anointed sites is inherently an ecological approach while the resolution of fundamental questions about biodiversity require answers grounded in a systematic biological approach" (Wheeler and Cracraft 1997, 439). What they mean here is that the biodiversity information question is not to be answered by surveying a few communities in depth but by doing the systematics work necessary to develop strategies for retaining the most genetic information. Ecological data has traditionally been collected at very small levels of scale—plots of $\leq 1\ m^2$ over relatively short periods of time (Michener et al. 1997, 330). It is not enough just to scale up and integrate over the multiple disciplines that might contribute biodiversity data (a difficult problem in its own right, as we have seen)—that data comes in two major incompatible flavors, with different scientific approaches both vying for the scarce resources to carry out data collection.

Biodiversity Science vs. Biodiversity Politics We have just seen that there is a problem with integrating data across a range of disciplines that have two fundamentally incommensurable ontologies. A second integration problem is that data that is collected is being integrated into two discourses—a scientific and a political discourse—which operate in two different (overlapping but sometimes analytically distinct) sets of relations.

A comparison of maps of systematics collections against species richness indicates one broad stroke of the problem: broadly speaking, species are a Third World commodity; information about species is a First World commodity (figures 4.9 and 4.10).

At the top level, there is a question of equity here. Underdeveloping countries (to use John Berger's phrase (Berger and Mohr 1975)) are becoming reluctant to share information and specimens, for good economic reasons. Thus Roblin (1997) bemoans the current difficulty of shifting microbial strains across borders: "The days when one could walk into any country with interesting habitats for microbial diversity and walk out with one's pockets full of interesting samples appear to be over. Countries containing such habitats now are aware that they may harbor microorganisms with commercial potential"

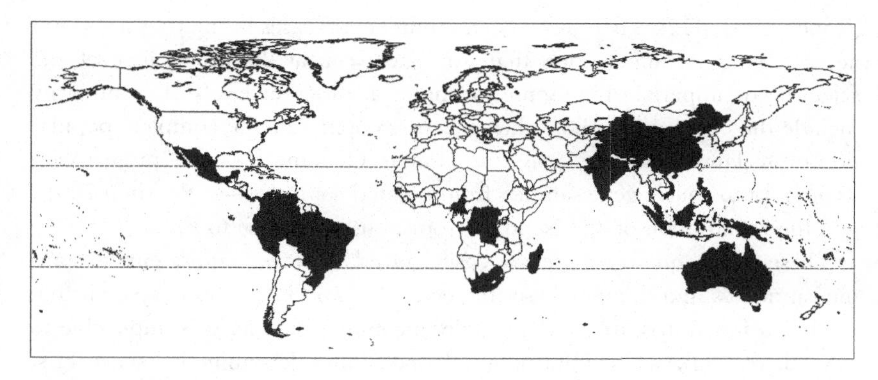

Figure 4.9
Countries with highest species diversity according to the WCMC National Biodiversity Index. *Source:* Brian Groombridge, ed., *Biodiversity Data Sourcebook,* WCMC Biodiversity Series No. 1 (1994): 142; http://www.unep-wcmc.org/protected_areas/world_heritage/reviews/forests/intro.htm~main.

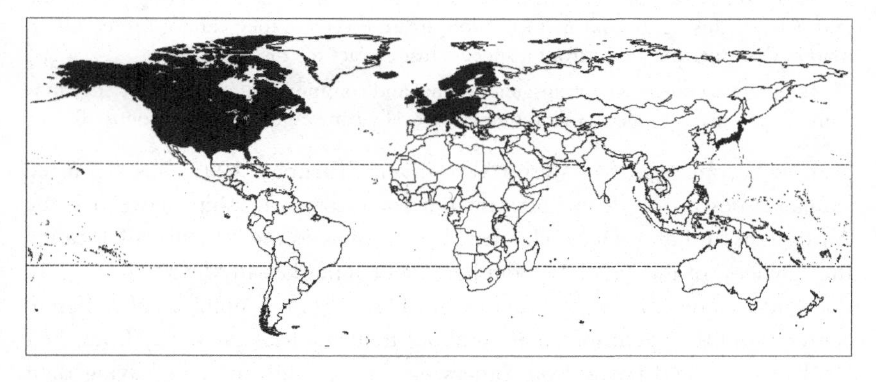

Figure 4.10
Countries with the most systematics collections in relation to national Biodiversity. *Source:* Brian Groombridge, ed., *Biodiversity Data Sourcebook,* WCMC Biodiversity Series No.1 (1994): 148; http://www.unep-wcmc.org/protected_areas/world_heritage/reviews/forests/intro.htm~main.

(Roblin 1997, 472). There have been innumerable cases in the past decade of access closing to information that was once seen at the prerogative of the scientific or imperial elite. Some examples at different levels of granularity include the controversy about that genetic sampling of the complete population of Iceland (Enserink 1998), the Human Genome Diversity project with its attempt to collect gene samples from isolated communities (Reardon 2001), and the repatriation of the Neodata tropical fish database to Brazil.

These equity questions are not easily solved, however, equity entails commensurability, and in many cases the economic and information systems that are being integrated are fundamentally incommensurable. It is impossible to give fair recompense for information if one cannot determine its owner. First of all, there is frequently a different understanding of "ownership,"as Posey (1997) writes, intellectual property rights

> (1) are intended to benefit society through the granting of exclusive rights to 'natural' and 'juridical' persons or individuals, not collective entities such as indigenous peoples. As the Bellagio Declaration puts it: Contemporary intellectual property law is constructed around the notion of the author as an individual, solitary and original creator, and it is for this figure that its protection are reserved . . . they cannot protect information that does not result from a specific historic act of 'discovery'.
>
> (2) Indigenous knowledge is transgenerational and community shared. Knowledge may come from ancestor spirits, vision quests, or orally transmitted lineage groups. (86)

Watson-Verran (e.g., in Watson-Verran and Turnbull 1995) has discussed similar issues arising from different ontologies of ownership between white Australians and aborigines. Hayden (1998) discusses the difficulty of locating the "owners" of information about herbs sold in markets in Mexico: frequently, the traders are peripatetic, buying the herbs from a number of different sources; and they gain information about their medical use partly from their local contacts and partly from others passing through markets buying their herbs and telling them their medical use. Who in this case should be reimbursed for giving Monsanto information about an herb that leads to lucrative drug development? In principle, there needs to be an ethnography of ownership prior to each particular determination; in practice, this just does not happen, and the requirement to respect intellectual property rights is honored by formally designating the trader as the font of knowledge. As Posey, Watson-Verran, and Hayden indicate, there is currently no standard, workable organizational interface permitting the fair exchange of information across cultural and economic divides. Even making the bold assumption of good will on all sides, then, there is continuing, de facto information imperialism, causing a net data drain out of the Third World into Western databanks. And yet information from many countries must be integrated in order to carry out biodiversity research and develop reasonable policies for the entire planet, since environmental questions as a whole do not respect national borders.

These equity issues speak to the difficulties of gathering together information for some notional database housed, more likely than not, in North America or Europe. Political and ontological questions do not go away once the hurdle of access is cleared. Indeed they are continually raised, through the multiple uses of biodiversity databases, in issues of data algorithms and granularity of descriptions. Edwards et al., for example, discuss the use of vegetation as a surrogate for animal species presence in Gap Analysis; they use vegetation because it can be classified from aerial photographs. In ground checks, they found that this led to more errors of commission then omission in the locating of animal species, but argued the following: "Given that Gap Analysis is a tool for predicting geographic distributions of terrestrial vertebrates for use in conservation planning, we argue that commission is preferred over omission" (Edwards et al. 1995, 4–10). Such generous errors are frequently made in estimations of the number of species in the world (Paul 1998, 3). Thus Stork (1997) argues that molecular-based species counts, which give higher estimates of numbers than counts using morphospecies concepts, are sometimes used as a political club (Stork 1997, 60)—and incidentally this contributes to the supplanting of morphology. Similar errors occur in estimates of the number of extinctions that are happening. This most certainly does not imply that biodiversity problems are not of crucial and pressing importance. To the contrary. It does, however, indicate the difficulties of using data in multiple ways. Relatedly, Klemm (1990) notes that one cannot legislate for the ways in which people will use the data in a public database; he describes one problem with contradictory implications as follows: "A difficult problem has always been to decide whether or not the location of endangered, rare or protected plants should be kept secret. Keeping the location secret avoids unscrupulous collection, vandalism or willful destruction by landowners fearing restrictions to development. On the other hand publicizing the location avoids inadvertent destruction in good faith" (28). Biodiversity work is integrally scientific and political, so it should not be surprising that its data shares the same features.

The Problems of Integration In this section, we have seen how ideally there are two kinds of integration going on in biodiversity work—between ecological and systematics data, and between scientific and political data. We have also seen that both kinds of integration cannot in principle be smoothly accomplished. Ecological and systematics data cannot be rendered equal just by standardizing over a set of weights and measures; and scientific data cannot be collected without making politically charged decisions. I have included both of these under the same general rubric because analytically many of the same processes are occurring in both cases: the forging of a dynamic uncompromise between agonistic groups in the very creation and structuring of biodiversity databases. The databases being developed today do not impose a hegemonic solution: they unfurl within them, at the level of data structure and data processing algorithms, the contradictions folded into their creation.

Time, Space, and Biodiversity Databases

I evoked earlier the problem of appropriate metadata standards for biodiversity databases. I have elaborated—through an analysis in turn of naming practices, context description, and information integration—three major broad dimensions of metadata information: how objects are named (and what is not named), how much information is given about data collection procedures and initial data use, and who are the intended (and unintended) users of the databases. On the one hand, I have argued throughout that what we need to know about data in a database is far more than the measurement standards that were used, and, on the other hand, I have argued that atomic elements of a database, such as measurement standards, contain complex histories folded (Deleuze 1996) into them, histories that must be understood if the data is to persist.

The organization of time, space, and matter is central to database construction. As discussed earlier, the earliest computer databases (see Khoshafian 1993) were *hierarchical*: one had a fixed hierarchy of data that one navigated sequentially in order to arrive at the node that housed one's information. In a hierarchical database, the data sits on the end of the nodes of the hierarchy so that if you change a branch, you have to move all the data hanging off that branch and sub-branches. This form of data organization has not gone away: there are still some botanical databases, organized around the backbone of a Linnaean hierarchy, whose designers have major problems incorporating changes in classification systems. The next major paradigm for database construction was the *relational* database, which in its ideal form (rarely realized in practice) organized data into a set of mutually exclusive tables of keys (record identifiers) and values (record contents); by manipulating sets of tables, one could access data nonsequentially. Furthermore, there was a degree of independence between data representation and data storage: if one wanted to change a hierarchical representation of the data one didn't have to move the data, just to produce a new table. Most biodiversity databases are relational, from Pankhurst's early European Taxonomic, Floristic and Biosystematic

Documentation System (ESFEDS) database for *Flora Europaea*, which used the Advance Revelation relational database system to, for example, such current databases as NAPRALERT—a database containing information from the literature on natural products (Gyllenhaal et al. 1993). The new generation of databases, as yet scarcely represented in the biodiversity world, are object-relational or object-oriented. These databases grew out of object-oriented languages, which in turn were conceived originally as attempts to represent the world as it actually works. The first object-oriented language was SIMULA (Khoshafian 1993), which attempted to simulate the way in which organizations carry out their tasks. The central insight was that one part of an organization did not need to legislate how another part did its job, it just had to provide that part with a set of inputs and be able to read the outputs. Object-oriented databases retain this characteristic: the user is shielded from the structure of the data. Thus any one object (and all its dependent objects) can be redefined without having to reorganize the whole. As far as databases of biological entities go, this means that the classification work that has to be "hard-wired" into a hierarchical system and built from tables in the relational model can be done on the fly at the time of the user's query in object-oriented databases. I have given this very cursory overview in order to make two points. First, the trend in databases has been to hold off on decisions about classifications until later and later in the information process. This represents a recognition of the fact that the bins one slots data into change dynamically over time. Second, and linked, the trend has been away from representing the world as a set of neatly nested entities to representing it as a set of autonomous entities (like Leibnitz's monads) that each may contain however many complex histories and processes, provided they can answer the queries made to them.

What can we say about these autonomous entities? The data format of a biodiversity database invariably reflects the set of temporalities and spatialities that are attached to the objects (taxa, species, genes, communities, etc.) being represented. We have seen previously how these objects themselves do not nest neatly inside each other along either of these axes. There is no clear absolute nesting of species inside genera inside families; it depends on what you call a species, and on what you do with border problems like hybrids, which do not fit neatly into any classification scheme. In a fully lawlike world, it would be possible to create a synoptic database that held all the information from all of these partial objects: we are begging the ontological question if we attempt to render our data fully lawlike.

Equally, there is no clear spatial or temporal nesting. We have seen that in the construction of biodiversity-relevant databases, there are two generally competing models: the ecological and the systematics; and that the ecological model is spatially organized and tries to break the world up into chunks (or partial objects, since they are only constituted as objects within a given discourse), which are appropriate units of biological activity. The appropriate temporality for any of these partial objects in space is singular. The systematics model is temporally organized and tries to represent the complete history of life on this planet—from the ticking of the genetic clock in mitochondria up to the enfolding of the history of all living beings in a coherent cladistic classification. One great divide temporally is between contingent historical time lines (stressing catastrophes like meteor showers, floods, etc.; cf. Ager 1993) and cyclical times (beloved of climatologists; see Lamb 1995). Objects in either of these temporalities can be unsettled by objects in the other. Thus there are cyclical stories about meteors (the movement of the solar system taking us into and out of dense patches on a cyclical basis) and secular stories about climate change (regular decrease in carbon dioxide in the atmosphere over time up to the present). Another temporal great divide, with equally complex sets of objects, is between stasis or equilibrium (the eternal present) and progressive change. A third temporal divide is between viewing human development as part of a natural series or of a sui generis social series. Thus paleontologist Elisabeth Vrba (1994) ties the development of a large brain in humans to heterochrony (the speeding up and slowing down of growth phases in the embryo) itself embedded in a pulse of global cooling, with large juvenilized bodies associated with cold temperatures and consequent vegetatively open environments (354). Anthropologists, on the contrary, argue about whether carnivory or eating tubers is at the base (Pennisi 1999). The natural scientist embeds a new temporality of development in a natural cycle (the Ice Age); the anthropologist delineates an archetypicaly social order (the cooked vs. the raw). It is an open question whether the two sets of registers and associated temporalities are contradictory: within biodiversity data there are many partially nested sets of such open questions—for example, geologists and climate scientists frequently see plate tectonics and the Milankovitch cycle trumping phylogenetic evolution, arguing that the latter is a possible process but is too slow to act before these larger changes winnows the set of species. The ontological question of whether rhythms and cycles or secular history and punctual events win out is integrally an eschatological one: every rhythm is nested in a catastrophe, and every catastrophe in a rhythm.

The point as far as databases go is that there is no uniform way of separating off the data objects (which themselves enfold complex histories) from their spatial and temporal packaging and inserting them in some other panoptical Cartesian space and time. As you nest cycles one inside the other, you find secular change irrupting into the story; as you nest secular narratives, cycles emerge. This is a problem across the divide. Furthermore, even within each divide, there is no simple nesting. The work of flattening out all the narrative sciences into a single narrative timeline is a *productive* effort that articulates data formats with relative power relationships between disciplines—through the mediation of classification systems and data standards. The manipulable second nature created within the computer is structured by an organizationally, politically, and morally inflected set of temporalities.

Similarly, you cannot just nest chunks of space one inside the other from the level of the entire globe down to the 1 m^2 patches of the traditional ecologist. A glance at figures 4.11 and 4.12 is emblematic of the high degree of variation of space at any level of scale—here, the whole world. In this example, a "standard" biogeographical model (figure 4.11) is pitted against a panbiogeographic model (figure 4.12). The panbiogeographic model comes out of a team based in New Zealand, an island deep in the Pacific, and it emphasizes the role of islands (by way, on recent theory, of terranes—pieces of continental plates that travel relatively independently, such as Point Reyes in California). The locus of object/space/time production is not an accidental feature here (cf. Bowker 1994, on geophysical theory and site). To the contrary, we have seen that the classification systems used tend to reflect their origins (e.g., the number of angiosperm species being an index of European plant life); in general in systematics, Heywood (1997) writes: "There are currently five or six different species concepts in use and no agreement between the different practitioners on how to develop a coherent theory of systemeatics at the species level. . . . In addition, species concepts differ from group to group and there are often national or regional differences in the way in which the species category is deployed" (9).

The same particularism can be noted temporally with respect to the presentism in much discourse about climate in climate change theory: the value that in many texts is placed on holding the world climate to current parameters. Climate change is the norm; the rate of climate change is historically highly variable; species survive. We seek to enforce an eternal present by congregating in low-lying areas in vast numbers, and then, like King Canute, ordering the seas not to rise. The importance of site indicates a central fact about biodiversity science: it is the science of the radically singular. The underlying question is what diversity there is in this world now, not what diversity there may be in earthlike planets under different sets of conditions. However, for many involved in biodiversity science, the totalization that is sought is a predictive, lawlike knowledge that will allow for an understanding of the wellsprings of biodiversity. To explore the difference between these positions, we

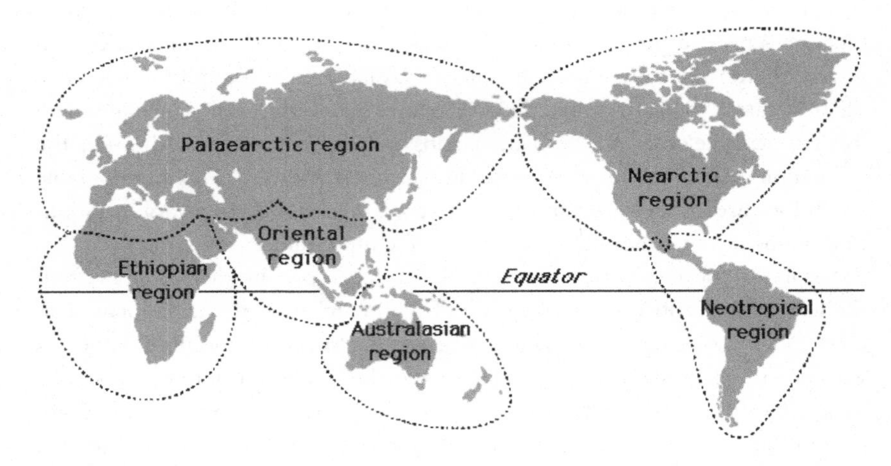

Figure 4.11
Biogeography: one way to slice up the world. *Source:* http://www.agen.ufl.edu/~chyn/age2062/lect/lect/lect_28/40_48.gif.

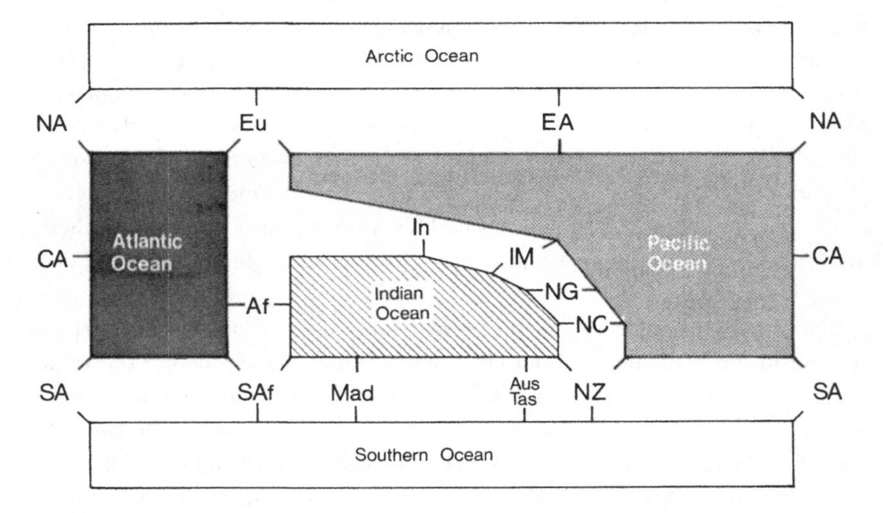

Figure 4.12
Panbiogeography: another way to slice up the world. *Source:* C. B. Cox, "From generalized tracks to ocean basins: How useful is panbiogeography?" *Journal of Biogeography* 25 (1998): 825.

will take a debate in TREE (Trends in Evolution and Ecology) in 1997. Kaustuv Roy and Mike Foote put forward the position that we should be measuring morphological diversity as well as phylogenetic diversity. What they meant by this was "some quantitative estimate of the empirical distribution of taxa in a multidimensional space (morphospace) that has axes that represent measures of morphology" (Roy and Foote 1997, 277). They claimed that morphological diversity was a more "natural" measure: people intuitively construct notions of diversity out of what looks different, not out of which genome contains more information (ibid.). Roy and Foote are defending a position of some weakness in recent systematics discourse: essentially, morphology is "out" and genealogy is "in"—the argument being that genealogy gives you causality, whereas morphology is an accidental (singular) feature. So they make an interesting rhetorical subsumption move, whereby it is phenetic variance in a morphospace that stands outside the flow of historical time and enters into scientific time: "The distribution of taxa in morphospaces reflects phenetic similarity (based on both primitive and derived characters), while the genealogy can be viewed as reflecting the 'routes of colonization' of that space (ibid., 280). Williams et al. (1997) respond by charging that the morphospaces give patterns, but that real, predictive processes that stand out of the particular will be found in genealogy: "What is at issue is not the universal superiority of one approach over the other, but the question of when to adopt one or the other. Descriptive measures of morphological diversity are needed by those exploring patterns in space and time. We believe that predictive measures are needed by those with more applied questions, such as how to choose areas to represent as much potentially useful biological variety as possible" (444). I am not interested here in the merits of the debate, but rather in two incidental features. First is the point that the argument for appropriate spatial and temporal measures is one that drives to the heart of a discipline: it is not an accidental feature that can be normalized in a notional panoptical biodiversity database.[8] The second is the resonance of the beautiful phrase "routes of colonization." This is a phrase that makes a point about genealogical systems in nature with

8. "Global synoptic or master species databases are being developed for a diverse array of groups such as virruses, bacteria, protists, fungi, molluscs, arthropods, vascular plants, fossils etc. and the SPECIES 2000 programme of IUBS, CODATA and IUMS has proposed that many of these should be joined to create a federated system that could eventually lead to the creation of a synoptic database of all the world's known orgnisms" (V. H. Heywood, "Needs for Stability of Nomenclature in Conservation," in *Improving the Stability of Names, Needs and Options: Proceedings of an International Symposium, Kew, 20–23 February 1991,* edited by D. L. Hawksworth (Königstein/Taunus, Germany: Koeltz Scientific, 1991), p. 10.)

respect to an abstract representational space. It is also a remarkably apt description of spaces and temporalities and objects of the biological sciences; they provide a trace of the radically singular routes of colonization of the world over the past several hundred years. It also operates at a meso-level: we can track the attempts to align various sets of objects with associated spatialities and temporalities across various sciences in terms of routes of colonization—whether this be physics colonizing geology or systematics morphology. . . . The abstract space and time of memory is unavoidably politically inflected.

The fossil record recapitulates all these problems. Koch (1988, 199) cites Ager on the taxonomic barrier whereby the Austro-Hungarian and British Empires can be traced through fossil collections in Vienna and London, respectively; this led to different sets of synonyms and thus to different apparent biogeographic regions. The political empire writ large on the natural world! Similarly, within the United States, he cites Sohl's analysis of apparent difference in species between the Texas and the Tennessee and Mississippi areas as being "artifacts of too stringent a taxonomy and an evident belief on the part of some workers that gastropods had very narrow dispersal limits'" (ibid., 199). Similarly, there are people—as Sohl points out—who have "'a tendency to either work on Cretacsous or on Tertiary assemblages but seldom both,'" thus emphasizing the differences between the eras; Erwin says the same for the Paleozoic and the Mesozoic (ibid., 199–200).

To summarize: each particular discipline associated with biodiversity has its own incompletely articulated series of objects. These objects each enfold an organizational history and subtend a particular temporality or spatiality. They frequently are incompletely articulated with other objects, temporalities, and spatialities—often legacy versions when drawing on non-proximate disciplines. All this sounds like an unholy mess if one wants to produce a consistent, long-term database of biodiversity relevant information the world over. At the very least, it suggests that global panoptica are not the way to go in biodiversity data!

Failure to name and standardize should not be read as a product of consistent contingent failures of nomenclatural bodies and data standards committees. On the contrary, it is deeply reflective of the nature of disciplinary research in biodiversity related sciences. These sciences deal in objects, spaces and times that cannot be readily normalized one against the other. To get an understanding of biodiversity data, we need to move away from a goal of producing a global panopticon, which will always be unattainable—the old chestnut of the incomplete utopian project (Gregory 2000). We should instead begin to look at the machines that are productive of local orderings and alignments of data sets. If we take a straight institutional reading of this process, then we see that it is reflective of various (sub)disciplines attempting to order the scientific world according to their own set of favored objects with attendant spatial and temporal units. The reductio ad absurdem of this position, as we have seen, is cybernetics posing as a universally valid language. However, we can also read the process in quite realistic terms: each (sub)discipline is acting as an effective spokesperson (Latour 1987) for the objects plus spatial and temporal units it produces. As a shorthand in what follows, I will call the objects plus spatial and temporal units "partial objects"—to borrow a phrase from Deleuze (1996)—"partial" to reflect the fact that these objects do not fully exist in the world and cannot be reconciled with any universal space and time, and yet "objects" to reflect that they have an autonomous existence. Ontologically, these partial objects can be read as trying to assert their own orderings of space, time and matter locally (and ultimately globally) in the world. In the agonistic world of science, the (sub)disciplines that speak for them are trying to do the same thing. Each is constantly faced with irruptions from other objects, which cut across their time lines, spatial units, or chunking of matter. Universal timelines just do not work, even though they are implicit in many of our data objects in the sciences of biodiversity (figure 4.13).

448498 BC 448497 BC 448496 BC 448495 BC 448494 BC 448493 BC 448492 BC 448491 BC
448488 BC 448487 BC 448486 BC 448485 BC 448484 BC 448483 BC 448482 BC 448481 BC
448478 BC 448477 BC 448476 BC 448475 BC 448474 BC 448473 BC 448472 BC 448471 BC
448468 BC 448467 BC 448466 BC 448465 BC 448464 BC 448463 BC 448462 BC 448461 BC
448458 BC 448457 BC 448456 BC 448455 BC 448454 BC 448453 BC 448452 BC 448451 BC
448448 BC 448447 BC 448446 BC 448445 BC 448444 BC 448443 BC 448442 BC 448441 BC
448438 BC 448437 BC 448436 BC 448435 BC 448434 BC 448433 BC 448432 BC 448431 BC
448428 BC 448427 BC 448426 BC 448425 BC 448424 BC 448423 BC 448422 BC 448421 BC
448418 BC 448417 BC 448416 BC 448415 BC 448414 BC 448413 BC 448412 BC 448411 BC
448408 BC 448407 BC 448406 BC 448405 BC 448404 BC 448403 BC 448402 BC 448401 BC

448398 BC 448397 BC 448396 BC 448395 BC 448394 BC 448393 BC 448392 BC 448391 BC
448388 BC 448387 BC 448386 BC 448385 BC 448384 BC 448383 BC 448382 BC 448381 BC
448378 BC 448377 BC 448376 BC 448375 BC 448374 BC 448373 BC 448372 BC 448371 BC
448368 BC 448367 BC 448366 BC 448365 BC 448364 BC 448363 BC 448362 BC 448361 BC
448358 BC 448357 BC 448356 BC 448355 BC 448354 BC 448353 BC 448352 BC 448351 BC
448348 BC 448347 BC 448346 BC 448345 BC 448344 BC 448343 BC 448342 BC 448341 BC
448338 BC 448337 BC 448336 BC 448335 BC 448334 BC 448333 BC 448332 BC 448331 BC
448328 BC 448327 BC 448326 BC 448325 BC 448324 BC 448323 BC 448322 BC 448321 BC
448318 BC 448317 BC 448316 BC 448315 BC 448314 BC 448313 BC 448312 BC 448311 BC
448308 BC 448307 BC 448306 BC 448305 BC 448304 BC 448303 BC 448302 BC 448301 BC

Figure 4.13
Booktime—from one million years—past 1970–1971; Volume 6: 498500 BC–398501 BC. *Source:* I. Schaffner et al., *Deep storage: Collecting, storing, and archiving in art* (Munich and New York: PresteSchaffner, 1998), 1101.

Evolutionary biology is tied to the "survival of the fittest" as a motor cause of biological adaptation: the mechanism may work in a normalized time and space, but not in the real world, where climatic events and tectonic shifts constantly upset the balance of the local environment and thus the effects of a given adaptation ("time" is not neutral). By the same logic, universal spaces do not work; there is no absolute set of "niches" in the world out there for organisms, since organisms themselves engineer their environments hugely in the long term.

Deleuze's analysis of Proust and signs describes the resonance between the activities of the partial objects and the activities of their spokespeople in terms of a resonance between the narrator (Proust) and his objects: be they characters (Charlus, Albertine) or materials (the paving stones, the Madeleine, and so forth). Deleuze argues that the singularity of Proust's vision is his argument that one cannot reduce the sets of data that he discusses to a set of laws (logos) true across space and time; rather, Proust denies any totalization, but seeks laws that can produce a local, contingent ordering of partial objects. In other words, Deleuze (1996) argues that Proust seeks to delineate sets of "machines" that, through an apprenticeship process, *produce* orderings of partial objects. I argue that this model provides a rich way of conceptualizing what is happening when we try to reconcile data across multiple disciplines in biodiversity research. Biodiversity science proceeds by conjuring partial objects into possible trajectories. This is its fundamental memory practice.

I have argued empirically for such a model in understanding biodiversity data. There are two immediate implications. First is that the resonance between narrator/narrative taken as the resonance between a scientific discipline and its objects is of fundamental importance for understanding the traffic in metaphors in science. It is no accident that a catastrophist geologist of the nineteenth century views the history of his own discipline in terms of catastrophic irruptions, or that a uniformitarian geologist sees his or her science unfurling henceforth in uniform fashion. Scientists in their narrative practice seek metaphors from their institutional practice, and vice versa. The second implication is that an observed convergence of a given set of sciences on a given set of objects, timelines, and spatial units is both an agonistic fact about the relationship between disciplines and a fact about the world. The convergence does not of itself entail the discovery of universal laws. To the contrary, if the history of science is any judge, it tends to operate as the convergence on a strategy for producing local orderings of partial objects: it is simultaneously the convergence on a given political economy and on a radically singular vision of the world. The machineries of orderings of partial objects (and the machineries of difference that drive them apart) are precisely where the rubber hits the road in the rendering of the world in the image of our globalizing ethnos.

I conclude this chapter with a more prosaic observation about biodiversity and its data. What we have seen is that the ordering of data across multiple disciplines is not simply a question of finding a commonly accepted set of spatial and temporal units and naming conventions, though this is the way that it is often portrayed in the literature (Michener et al. 1998; Michener et al. 1997; Dempsey and Heery 1998). To the contrary, these ordering issues lead us very quickly on the one hand into deep historiographical questions and on the other to questions of institutional politics. If we are going to develop decent biodiversity policies, then we need databases held together through good metadata practice; and in order to work out the latter, we need to recognize the depth of the problem. In a biodiverse world, we need to be able to manipulate ontologically diverse data.

5

The Local Knowledge of a Globalizing Ethnos

Religions have no existence outside of our ideas and feelings, whereas the social and political system subsist in the order of things, which is the source of our ideas and induces our feelings.

—J. Benner, *Commentaire philosophique et politique sur l'histoire et les révolutions de France, de 1789 à 1830*

In subjects of pure Science an Individual can speak as if he were all Mankind: for tho' it is he who speaks, yet that which is spoken is the Reason, which is *one* in all men. From whatever point of the periphery, or of its Area, the Word proceeds, in the moment of its utterance it becomes the same everywhere within the sphere of its audibility.

—Samuel Taylor Coleridge, "Notebooks"

A government seeking to entrench passive obedience desires that education be limited to abstract science; thus under the empire and since 1815, only exact sciences have been taught at the Ecole [Polytechnique]. A new order of things calls forth other principles.

—M. Doré, *De la nécessité et des moyens d'ouvrir de nouvelles carrières pour le placement des élèves de l'Ecole polytechnique, et de l'utilité de créer de nouvelles chaires dans cet établissement*

Just who we universals are is not an obvious question. The Athenians grounded their equality in their autochtony (Loraux 1996): it was a foundation of their equality, a foundation that excluded women (not born of the earth) and outsiders. The phrase is resonant today: the French with their *terroirs* at home and universalizing French empire abroad still refer to themselves as "autochtones" (Détienne 2003). The tension between universality and rooted locality runs deep. Possibly closest to our globalizing ethnicity (I speak for my fellow uni-

Epigraphs: J. Benner, *Commentaire philosophique et politique sur l'histoire et les révolutions de France, de 1789 à 1830* (Paris: Treuttel et Wurtz, 1835), p. 19; S. T. Coleridge, "Notebooks," in *British Museum Add. Mss. 47546* (London: n.d.), f. 17; M. Doré, *De la nécessité et des moyens d'ouvrir de nouvelles carrières pour le placement des élèves de l'Ecole polytechnique, et de l'utilité de créer de nouvelles chaires dans cet établissement* (Paris: Bachelier, 1830), n.p.

versals) is that of the Goths: they have an unfortunate reputation vested on them by their victors as ravaging hordes, but they are arguably the first political group for which *ethnicity* was not an issue: one became a Goth through ascribing to the Gothic political organization (Wolfram 1988), just as one becomes a member of Axis of Good through ascribing to a given form of political organization that goes under the universalizing name of democracy. We universals are the only group to have a nonlocal history—the truths we discover were always already there; they are not rooted in a creation myth tied to this spring or that species. We are the group that has a maximal break with the past—as Michelet taught us, we are not determined by our climate (others are); as Comte taught us, our knowledge is not tied to its roots in metaphysics and religion but is increasingly liberated from same (unlike the knowledge of others). We are the group that maximally rejects its past except as pure flow into an ideal eternal present, the real present that lurks beneath the appearance of disorder.

In this chapter, I explore the role of the coin and the list as tokens of our universal nature that order our memory practices in such a way as to create a manipulable second nature. In so doing, I examine what is particular and irreducibly local about our universalizing way of remembering and recounting the past.

Two Sides of a Coin: The Canonical Globalizing Modalities of Memory Practice

Biodiversity is the feel-bad word for the new millennium. We all know that we want it and that there is a lot less of it around than there used to be. Indeed, as a species, we have irrupted into timelines stretching back some 700 million years as the cause of the sixth greatest extinction event in the history of the earth. Extending into the far reaches of Braudel's *longue durée* (1973), we are up there with the meteor that (possibly) killed the dinosaurs. We operate truly globally—affecting every nook and cranny on the planet—except, perhaps, the huge frozen lake Vostok in the Antarctic, which "is absolutely devoid of interference. The youngest water in it is 400,000 years old. It doesn't know anything of human beings, fossil fuels, or plastics. It is a window into life forms and climates of primordial eras" (http://www.earth.columbia.edu/news/story03_21_02.html). And even this we are working to explore, at first noninvasively, with "radar sounding, laser altimetry, magnetics, and gravity surveys," and at the end of the day we will probably send a putatively clean robot down there (Bell et al. 2002) (see figure 5.1). We commandeer an astonishing percentage of the sun's energy stored on earth, the freshwater that sculpts

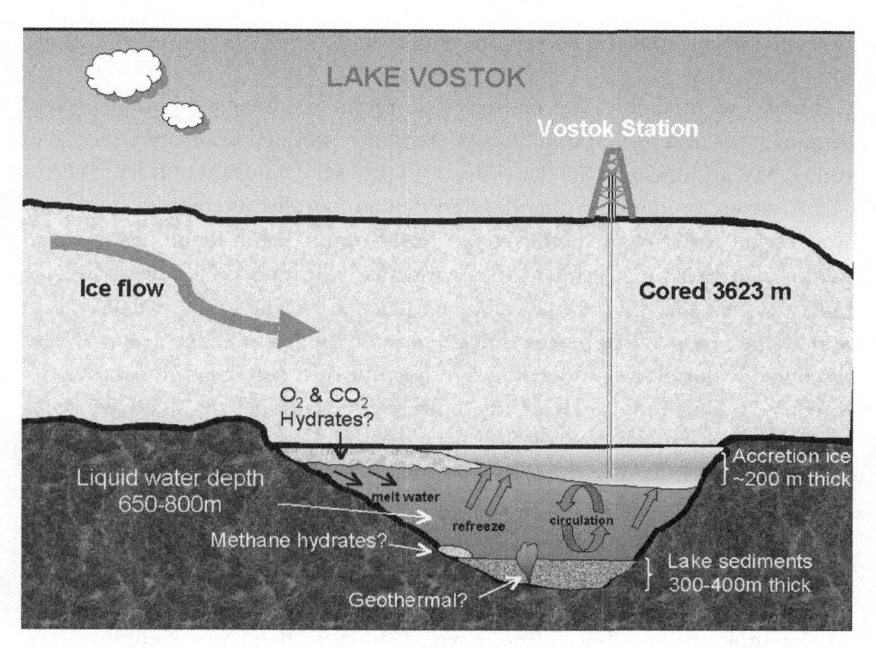

Figure 5.1
Almost there (the act of penetration). *Source:* http://www.homepage.montana .edu/~lkbonney/IMAGES/Presentations/AAAS%20Feb%2001/Lake%20Vostok%20 Schematic.jpg.

its features, and are even digging into the earth's archives to release trapped energy in the form of petroleum.

There are two dramatically different modalities for dealing with the question of biodiversity. In the first, one tries to accord every category of living thing a single biodiversity value, so that the policymakers can start the work of determining what should be protected and what should not—in much the same way as we now internationally barter pollution. Drawing on Donna Haraway's work (1997), I call this a modality of implosion. In the second, one tries to list every last living thing—a frenzy of naming that is reaching its apogee with several multi-million-dollar international efforts to record just what there is out there. I call this a modality of particularity.

The two modalities immediately call to mind two great creations of bureaucracy: the coin of the realm (which Schmandt-Besserat (1992) places at the origin of writing) and the list (which Jack Goody (1986) places at the origin of writing). Each are learned responses developed over millennia to deal with complexity and scope—how to handle a large-scale enterprise through abstraction and classification. So it's not surprising that these two behemoths are stalking biodiversity.

These two modalities are constructed around a similar temporality: background stasis and foreground change (as in the production of animation pictures). My argument is that in order to write our "natural contract" (Serres 1990), we are producing a singular and rich temporality as complex in its own way as that read out of myths by, say, Lévi-Strauss. It is a temporality that in addition to being powerful in the world (for who can doubt the power of bureaucracies and the efficacy of technoscience?) is integrally eschatological and mythic. I argue that paying due attention to the full richness of our current discourse about biodiversity entails reading our own emergent global society's discourse just as we would read any other discourse in societies that have never been modern (Latour 1993b).

The argument comes in three parts. First, I give a brief account of some recent work in the history of money, as a way of opening up the issue of what we can look for in the modalities of accounting for biodiversity that will be the topic of my inquiry. Second, I look at one organization of the modality of implosion. Third, I look at one organization of the modality of particularity. My examples will be drawn from current efforts to database—figuratively, or as we will see literally—life on earth.

Money, Memory, and Discourse

The archetypal figure adjudicating between boundaries is that of the merchant, who is the trader between the inside (members of the *polis* in classical Greece) and the outside (neighboring communities that wish to operate some kind of trade). In his book *Le Prix de La Verité* (the price/prize of truth), Marcel Hénaff traces the vicissitudes of this mediation over time, tying it initially to Plato's apothegms against the Sophists selling that which is beyond price—philosophical truth. At this historic conjuncture when we are renegotiating the natural contract, the array of characters may be different, but the figure remains the same. We have something beyond price: the miraculous bounty of the earth—that gift we enjoy or invaluable creation we steward (Worster 1994). We are in the process of setting a price on it. The boundary between nature and culture we are creating is similarly textured to that of the definition of a community; it can be strictly geographical (natural wilderness on one side and urban mean streets on the other) but is more generally the outcome of heterogeneous, partly conflicting operational definitions. The merchant figures in this case are international organizations like the Organization for Economic Cooperation and Development (OECD), which has in the past decade taken up the banner of brokering international deals on the environment:

> A healthy environment is a pre-requisite for a strong and healthy economy, and both are needed for sustainable development. The OECD provides a forum for countries to share their experiences and to develop concrete recommendations for the development and implementation of policies that can address environmental problems in an effective and economically efficient way. . . . Increasingly the problems they face are more complex, and will require co-operative action at the international level (e.g. climate change) or coordinated packages of policies across regions and/or sectors (e.g. biodiversity, agricultural pollution, and transport). . . . OECD supports its governments in addressing these problems primarily through the work of its Environment Policy Committee, through Joint Working Parties on Agriculture and Environment and on Trade and Environment and through Joint Meetings of Tax and Environment Experts. Overall, these activities contribute to the crosscutting work of the OECD on sustainable development. (http://www.sourceoecd.com/content/templates/co/co_main.htm?comm=environm)

Sustainable development marches under the proud banner of OECD's programmatic definition, which has been endorsed by the World Bank, the International Monetary Fund, the World Wildlife Fund, and the World Heritage, among others, and was common to participants in the World Summit on Sustainable Development held in Johannisburg in 2002. Contemporary discourse of biodiversity is structured within this policy framework (Takacs 1996).

In order to render two things (species, wetlands, pollutants) comparable, one needs a token that can circulate in their stead—so that, for example, you can trade a marine habitat with such and such a degree of richness for a wetland area with a comparable degree. Not even the spatiotemporal unit comes prepackaged. In order to preserve a wetland, you have to preserve its adjacent water table, since draining an adjacent field can drain the wetland indirectly. Similarly for time: if you look at the decimation of caribou stock through current logging practices, you get a very different picture of sustainability if you take the base unit to be ten or two hundred years (Jonasse 2001). Yet you do need a tradable unit that can circulate freely without containing too much historical baggage. In *Genesis*, Michel Serres (1982a) writes that money is the "degree zero of information." He argues that in order to render a thing (a commodity) or an action (digging a field) into money, then all detail about the nature of the thing or action has to be blackboxed, so that what is left is the smooth surface of a coin or note, with a quantitative value attached to it. Money, by this account, constitutes the least possible information that can be shared about events and objects while still maintaining a viable discourse around them.

When he refers to money as a degree zero, Serres is not asserting that it is an empty set. Indeed not. As Hart (1999) remarks:

> The word *money*, as I mentioned at the beginning, comes from Moneta, a name by which the Roman queen of the gods, Juno, was known. . . . Moneta was a translation of the Greek Mnemosyne, the goddess of memory and mother of the Muses, each of whom presided over one of the nine arts and sciences. Moneta in turn was clearly derived from the Latin verb *moneo*, whose first meaning is 'to remind, put in mind of, bring to one's recollection'. . . . There seems little doubt that, for the Romans at least, money in the form of coinage was an instrument of collective memory that needed divine protection, like the arts. (256–257)

What is remembered in the coin is precisely that which is needed in order to carry on economic discourse. In so doing, the coin continually evokes (recalls) the compact made with the state to honor information about value expressed in the form of an amount on a coin. Hénaff traces the etymology of *alatheia* (the female avatar of truth in Attic Greece) to "memory" as well. The struggle between the "Sophists" and the "philosophers" about selling truth mirrors that between the global policymakers and deep environmentalists about dealing in biodiversity. The philosophers and environmentalists go for deep, "real" truth or wilderness (total memory of thought or world), the Sophists and policymakers for marketable truth (a minimal memory set).

The trouble is that while it is clear in a general sense that we as a globalized species and globalizing economy are currently deeply renegotiating the relationship between nature and culture, we really have no place to site a reflective discourse about the range of ecological and economic issues. We can take a lead from Lesley Kurke's *Coins, Bodies, Games and Gold* (1999), in which she explores the discursive dimension of money through analyzing texts in Herodotus. She starts from the curious fact that there is exceedingly little mention of coins in texts for the first two hundred years after the first minting. Indeed, one must wait until Aristotle to find a philosophical treatment of money. However, she notes, there is a rich thread tying together two alternative modes of discourse. The first is the discourse of the symposium. This is associated with leisure, aristocracy, the masculine ideal, and pure metals (gold and silver). Then there is the discourse of the agora, associated with bustling labor, merchants, effeminacy, and base metals (alloys standing as surrogates for the coin). I locate a similarly rich and heterogeneous list associated with the language of developing a single currency for biodiversity; figuring prominently in that list is a reading of social time and of memory—here I concentrate on social time.

Modalities of Implosion: The Language of Money in Biodiversity Discourse

The word *biodiversity* is of relatively recent coinage; it is no more than forty years old. It was developed within the emerging field of conservation biology—

a science with a mission (Takacs 1996) to preserve our ecology. Nils Eldredge (1992) argues that the ecological perspective (as opposed to the taxonomic perspective, which I look at in the next section) is an economic one—talking as it does about the way in which species interact in the economy of nature. It thus harks back to the common roots of ecology and economics in the Greek word for household, *oikos* (Williams 1983).

It has been seen as increasingly important within the biodiversity community to bring ecology and economy together, to find a way of expressing the value of "ecosystem goods and services" for humanity. The argument goes that biodiversity conservation can only take place if we have a powerful language shared between scientists (who often see themselves as philosophers who have been forced into Sophism) and policymakers. The former wants to pack as much complexity as they can into the token that policymakers can then exchange, without knowing anything of the science. Just as the customers don't want to know details about the labor and art that went into forging a bust, they want to know just enough so that they can be assured that the outlay of money is reasonable. Thus the policymakers can say: "We will take this bit of wilderness but we will give you another bit, which has, in the best of all possible worlds, an equal or greater biodiversity value. Or we can lose that species if we preserve another of similar value." Only if we can account for diversity will we be able to preserve it.

So how does one go about measuring biodiversity? The intuitive step of assigning a unit value to each species and then totaling species counts in a given area will not work for two reasons. First, you want maximum *spread* of biodiversity value if you want to save a useful minimum set for life on earth: "For example, a dandelion and a giant redwood can be seen to represent a richer collection of characters in total, and so greater diversity value, than another pair of more similar species, a dandelion and a daisy. . . . This shows how the phenotypic characters (or the genes that code for them) could provide a 'currency' of value for biodiversity. Pursuing this idea, we will then need to maximize richness in the character currency within the conservationists' 'bank' of managed or protected areas" (Museum of Natural History 2002). In other words, there's no point in preserving a large number of species within a small spread of genetic difference. Second, there's no way to preserve just one species, so it is not a useful unit of analysis:

> Often, higher-order species on the food chain have the most exacting environmental requirements and are thus valuable indicators of the health of the entire ecosystem; they or others may be critical 'keystone' species because they are located at the center of a network of interdependencies. Thus, as a practical matter, species values become proxies for ecosystem values: the Endangered Species Act in the United States is an embodiment of this principle in policy. And of course we regularly justify large expen-

ditures to save some species (e.g., the African rhinoceros) but not others (there is no Save the Furbish Lousewort Society) (Goulder and Kennedy 1997, 36)

Some species, then, are more important than others, since they stand as proxies for ecosystems. Species congregate in complex ecological groups.

The species, which is the proximate unit we most intuitively respond to (Lakoff 1987; Stevens and Cullen 1990), holds then a tension between information (going down to the gene level) and community (going up to the ecosystem level). Further, the species concept is of itself highly controversial; there are a number of conflicting ways of severing the great chain of being (Wilson 1999). Central here is that species are not stable, well-defined entities. The difficulty is in trying to partition emergent processes into stable analytic coins. Much biodiversity discourse centers on preserving that which is—a current set of species, our current climate conditions, and so forth. We talk about preservation and conservation, not potentiating dynamic change. On this logic, we should be preventing orogenesis, which has a huge impact on climate change—the thrust of the Indian subcontinent into Asia that is throwing up the Himalayas has been a significant cause of the lowering of temperature by trapping carbon dioxide; further, volcanic outgassing is seen as a major variable in lowering the temperature by causing higher reflection of the sun's energy (Huggett 1997). Paradoxically, preventing global warming is extremely harmful for biodiversity—when there were temperate forests up in the Arctic, the biodiversity potential of the world was higher than it now is. A second paradox of the battle between saving stable sets and potentiating change is that it leads to preserving ecosystems that today might seem particularly uninteresting to those who care about the environment. Thus Terry Erwin and others talk of preserving evolutionary potential, or "species-dynamo" areas:

However, there are great difficulties in predicting future patterns of diversification . . . in patchy and changing environments, particularly as projected human-driven changes are unlikely to reflect simply those of the past. Following Erwin and Brooks et al.'s arguments, the perverse result of extrapolating future diversification 'potential' from recent history is that it leads to favoring conservation of species that are particularly similar to another (e.g. faunas with large numbers of rodents), in preference to biotas with more dissimilar and diverse species. (Williams et al. 1994, 71).

The projects of preserving the possibility of change for a rich future or preserving the current set of species are at best kissing cousins.

Most units defined analytically in conservation biology run into this problem. If you go up a level to the ecosystem, you run into the problem of defining just what sort of a thing an ecosystem is. R. V. O'Neill (2001), for example, argues that the ecosystem concept stabilizes the system "at a relatively constant equilibrium point"; indeed, he goes on to say: "Concepts like stability and ecosystem are ambiguous and defined in contradictory ways. In fact there is no such thing as an integrated, equilibrial, homeostatic ecosystem: It is a myth!" (3276). The word "myth" is a useful one here, since it is surely what we are dealing with. Much biodiversity currency discourse is concerned with rendering the present eternal—moving ourselves and our planet out of the flow of history. We want this set of species to last; we want this climate to continue, and so forth. The background (our canvas) should stay stable while the foreground (human attainment of perfection) should be changing rapidly—even if we no longer use the term in vogue from the 1830s to the 1960s in the West: progress. The *nec plus ultra* is the cloning movement. Thus a company in San Diego offers gene banking by holding out the possibility of pet cloning (figure 5.2). Indeed, one vision (popularized in the film *Jurassic Park*) is that we can preserve biodiversity by banking gene sequences and rolling out diversity when we need it.

Here is one of the central problems of trying to collapse multiple registers into a single currency containing just the necessary information; the resulting units of analysis will be riven with contradictions: "If there is no stable equilibrium, why bother to conserve? . . . How do you restore ecosystems when you don't know what to restore them to?" (O'Neill 2001, 3276). Reid Helford (1999) has written about this difficulty within the oak savannah restoration project in Illinois. He points to the difficulty of deciding what is natural and what is human (and indeed to engineering a division between the two). The restoration project is trying to restore the ecosystem as it was before European settlement. And yet the Native Americans—through fire technology (Pyne 1997)—were central in the creation of that ecosystem. Given the relatively recent orogenesis of the Rockies, the new prairie ecosystem turned duocrop (except along train lines) of the American Midwest has been created out of a string of invasive plants, animals, and people. For the policymakers in the project, white humans fall on the side of culture and so are external to the ecosystem, whereas Native Americans fall on the side of nature and are internal to it. O'Neill (2001) points out that in general we constitute "the only important species that is considered external from its ecosystem, deriving goods and services rather than participating in ecosystem dynamics" (3277; cf. Eldredge 1995). That old nature/culture divide—so central to Lévi-Strauss's mythologies—is alive and well and equally torqued within modern technoscientific mythology.

Which brings us to the question of the work that is being done in order to effect that divide today. The mode that I will examine here is the move to value "ecosystem goods and services,"—a move that has structured the discourse of

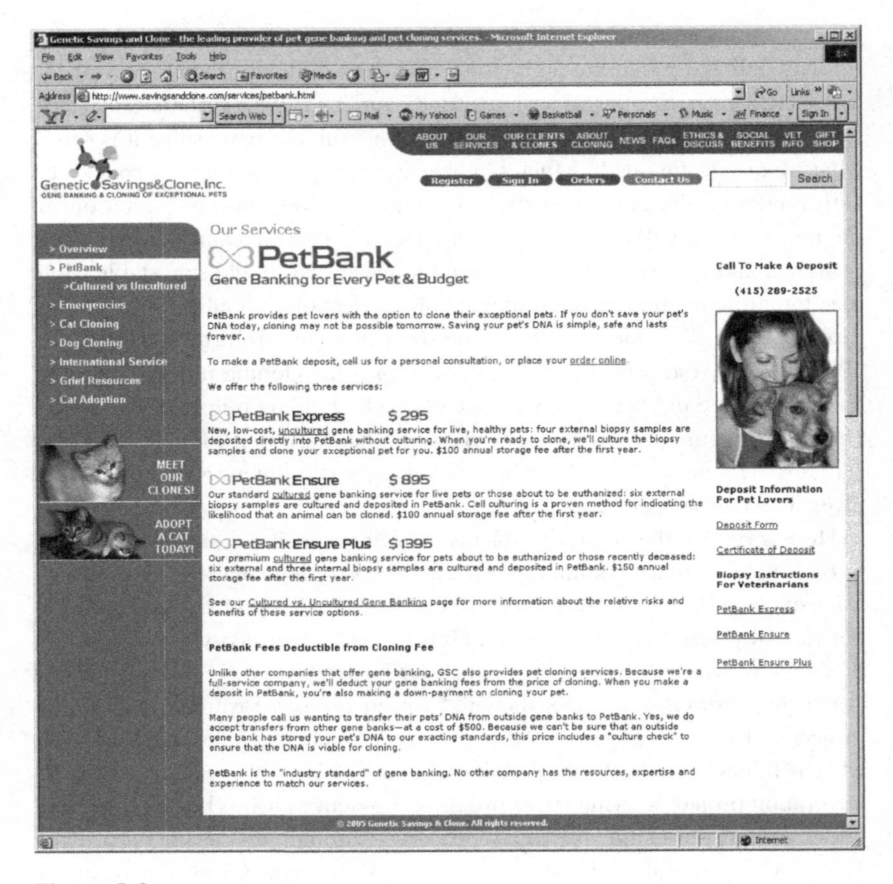

Figure 5.2
Banking those genes. *Source:* http://www.savingsandclone.com/services/petbank.html.

seeking to value biodiversity. Now if we are external to nature, we stand in the position of the Creator—outside the flow of history, acting on, but not being of, the world. It is possible to argue that this move's current form is associated with the science of the Industrial Revolution and, in particular, Lyell's geology (which discusses the question of man and nature at length).

So. Nature is external. We need to find a way to express nature in terms of our own value systems. If we start putting number values on aspects of our environment, we quickly run into the problem of infinity: "As a whole, ecosystem service have infinite use value because human life could not be sustained without them. The evaluation of the tradeoffs currently facing society, however, requires estimating the *marginal* value of ecosystem services (the value yielded by an additional unit of the service, all else held constant) to determine the costs of losing—or the benefits of preserving—a given amount or quality of services" (Daily 1997, 8). And there is a lot of infinity about. Take the soil, for example: "Soil provides an array of ecosystem services that are so fundamental to life that their total value could only be expressed as infinite. . . . Human well-being can be maintained and fostered only if earth's soil resources are as well." (Daily et al. 1997, 128). Now infinite values are not of much use in economics, so in general the shift is toward dealing with units of analysis that produce finite numbers and delete inconvenient infinities (a ploy borrowed from the physics community perhaps).

Accordingly there has been an attempt to separate analytically the different kinds of value that nature provides. Use value is our current use of the ecosphere. This produces a very large but vaguely quantifiable number: "Despite recent estimates that the Earth's ecological systems are worth about $33 trillion annually, the comparatively low cost of maintaining the biological diversity that underpins these services is ignored" (James et al. 1999, 323). Not even Bill Gates can rival the ecosystem; however, he may well be worth Australia. This use value can be taken at any unit of analysis—the ecosystem, the species, or the germplasm: "In common with all agricultural crops, the productivity of modern wheat and corn is sustained through constant infusions of fresh germplasm with its hereditary characteristics. . . . Thanks to this regular 'topping up' of the genetic or hereditary constitution of the United States' main crops, the Department of Agriculture estimates that germplasm contributions lead to increases in productivity that average around 1 percent annually, with a farm-gate value that now tops $1 billion" (Myers 1997, 257). This last is from a paper describing our genetic "library"—the modern form of the book of nature metaphor that has a history stretching back many centuries. The figures given are frankly absurd—there is no way that such measures can be made of a system we are part of and in a world in which we don't build statistics in such a way that they could possibly reflect use value—but at least they are finite. We are moving into the implosion of multiple registers into a single value.

But use value alone is not enough to describe the value of biodiversity; there is also option value. Option value is the interest that we have in keeping our current stock of biodiversity against possible future uses. Thus a rare strain of corn in Mexico might help us if a new parasite emerges that attacks all other strains of corn apart from this one. . . . Infinity rears its ugly head again; option value with genetic features as the basic currency "gives any included attributes equal value because of the inevitable ignorance or uncertainty of precise future needs. Biodiversity conservation would then focus on maximizing the amount of 'currency' (the number of valued biological attributes, features or characters) to be held within the protection system 'bank' (the set of protected species, ecosystems or areas). Thus the paradoxical consequence of equal value for attributes as units of currency is that their owners, the individuals, species or areas, may have different values because they contribute different numbers of complementary attributes for representation in the protection system" (Williams et al. 1994, 68). A third major category added to option value is existence value—the value that I derive from the existence of the Grand Canyon, say, even though I have no intention of ever going there (Goulder and Kennedy 1997; Humphries et al. 1995, 102–103). I feel that way about Mount Uluru (the rock formerly known as Ayer's). This value is rarely quantified.

With use value and option value as the key components of a biodiversity currency, being those components that produce numbers, they are expected to do a lot of work. They should stand as proxies for other measures, which get a mention but are then to be ignored. The World Conservation Union stated that "the justification for preserving genetic diversity is that it is 'necessary to sustain and improve agricultural, forestry and fisheries production, to keep open future options, as a buffer against harmful environmental change, and as the raw material for much scientific and industrial innovation—and as a matter of moral principal" (Williams et al. 1994, 68). The potentially infinite—the moral principle in this case—gets pushed to the sidelines, with the unstated assertion that moral principle will be served by maximizing use and option value. Gretchen Daily (1997) makes the same move, relegating the infinite to the sidelines and structuring the economic argument in such a way that its value is incorporated: "Our concentration is on use values; aesthetic and spiritual values associated with ecosystem services are only lightly touched upon in this book, having been eloquently described elsewhere" (Daily 1997, 7). However, surely Derrida gets it right: that which is excluded is often that which structures the discourse.

The valuation is created as a way of collapsing multiple registers (aesthetic, religious, spiritual) onto an artificially created unit of currency that can then circulate within modes of discourse hostile to just these sets of registers. In a sense, it's surrogacy all the way down in biodiversity research; the only way you can measure all the life in a given area is to follow Terry Erwin's model, say, and fog the area to count dead beetles: an efficient mode of counting that has the downside of possibly destroying some highly specialized species (beetles can be specific to a given tree). All Taxa Biodiversity Inventories are slow, clumsy, and very expensive. All else is surrogacy, such as the aerial map of vegetation cover standing as a surrogate for animal life with the assumption that we know which species tend to be associated with which cover. Williams et al.

(1994) start their paper on biodiversity by sidelining the difficult issue of surrogacy when trying to implode multiple values (here "the aims of conservationists"): "We ask whether maximizing inter-specific genetic diversity necessarily fulfils the aims of conservationists most directly, or whether the consequences of this choice may actually be at odds with their objectives, and whether more appropriate currencies for conservation can be identified. We are not concerned here with the extent to which one currency can serve as a surrogate for another, which is regarded as a separate issue" (67). They conclude by embracing it: "In reality, currencies may yet prove to be highly correlated among species, so that any direct diversity measurement could present an approximate surrogate for any other, although this remains to be confirmed" (ibid., 76). The currency, then, holds out the promise of collapse of multiple social values onto a single measure. If engineered correctly, this currency will enter into policy discourse in just such a way as to promote a broadly common set of values held by conservation biologists. This is a dangerous move, akin to one studied by Bowker and Star (1999) by nurses seeking recognition for their *process* work by cutting it up into regular temporal units (half-hour work units) which could then be recognized within hospital accounting systems. In the case of biodiversity, the currency move is collapsing emergence into units (the commodity form) that circulate in a very flat, linear time and space (Sohn-Rethel 1975).

The money tokens that are created must stay in circulation and in a space that has been evacuated of events. The eye should be on foreground change (human development, or defining the boundary of culture) against a persistent canvas (background stasis, or defining the boundary of nature).

Modalities of Particularity: The Tree of Life

A second modality for accounting for biodiversity is that of the tree of life (figure 5.3). This is the art of the particular—any surrogate is a counterfeit, and to counterfeit is death, as the original "Ben Franklins" said. The tree of life is a venerable mode of representation of our knowledge about life, its origins and development. Life starts at the root, the single-celled protoplasm, then claws its way up the tree until it pinnacles at Homer Simpson. This representation is an extremely powerful one that stems from an unsystematic but very general move in the nineteenth century away from classifying objects by their innate qualities (the Aristotelian turn) to classifying them by their genesis (Tort 1989). This new classification modality was associated in complex ways with the regime of governmentality (Foucault 1991). The emergent technoscientific empires of the nineteenth century developed the discipline of statistics (which etymologically refers to the State) and new systems of classification

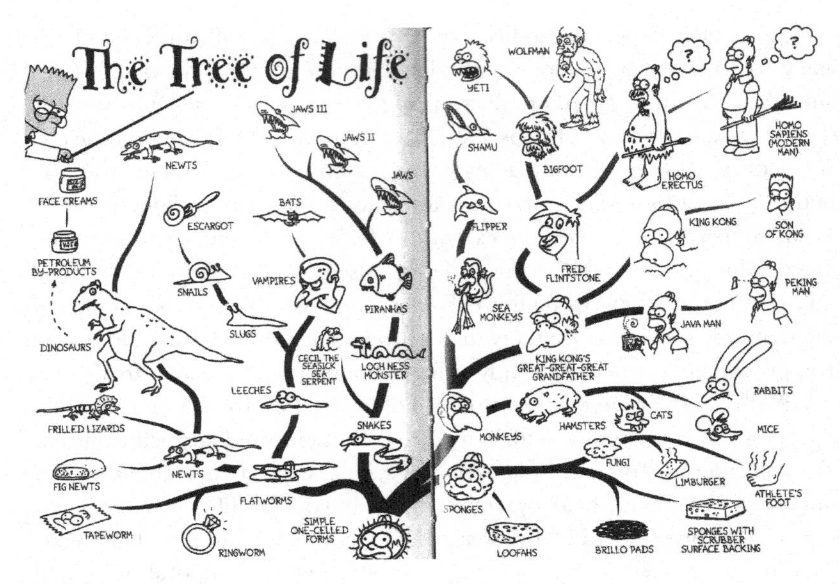

Figure 5.3
The traditional tree of life. *Source: Bart Simpson's Guide to Life.*

to deal with the vast amounts of impersonal information that had to be collected in order for the Empire to function efficiently. Today, in both the social and the natural sciences, the tree is starting to look somewhat ragged. A more modern form of the tree is less attached to its roots (figure 5.4). This new tree has no clear roots in the ground; it's an exercise to work out where the origin is stashed away. There's a rough red circle around viruses. These entities don't have simple genetic histories the way that larger organisms do: they are sometimes seen as being devolved from higher life forms (a parasite, that prototypical troubler of inside and outside both physiologically and socially (Serres 1982b), which discovered a simpler way to get its genetic message across) or as evolving in pace with their host rather than from any internal mechanism. This is the problem of Occam's razor. It is a computationally huge task to calculate all the possible phylogenies (branches of the tree of life); many possible routes lead to the present. Thus when producing computer models, we assume that time is unidirectional: species cannot lose characteristics once acquired. And yet we know empirically that some species do just that. It is assumed that history is simple. One species can never branch off more than one species at a time. There's no particular reason for this assumption other than that it makes the calculation possible with current technology (and this is perhaps reason enough . . .). It is assumed that this simple history only has one under-

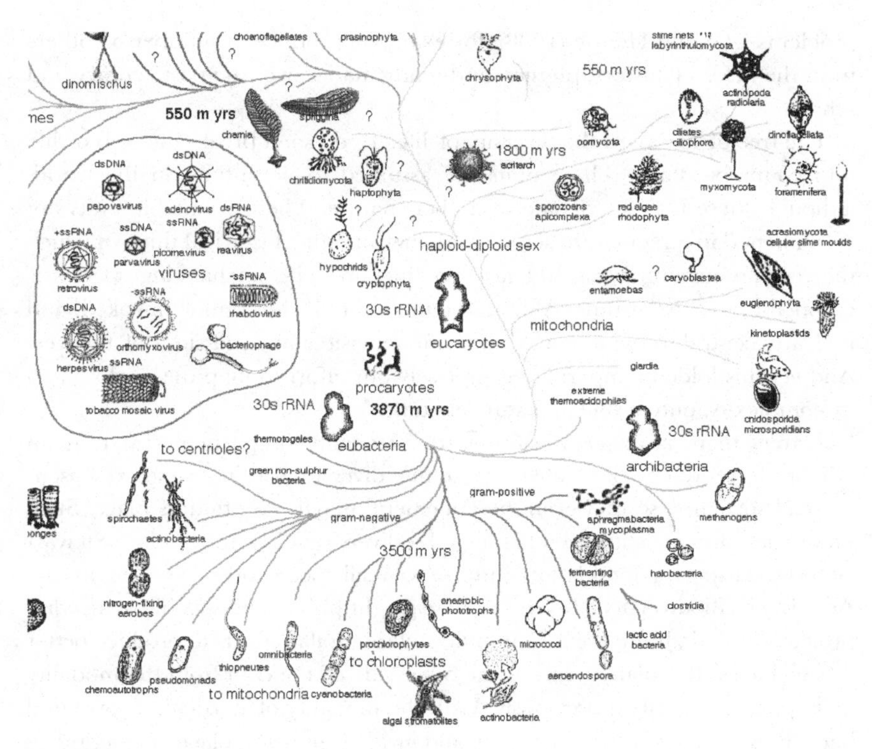

Figure 5.4
The tortured tree. *Source:* http://www.dhushara.com/book/evol/trevol.jpg.

lying cause. If genes can spread by contagion rather than be adopted from parents, then the problem of calculation becomes truly staggering. And yet we know that some genes spread by contagion (see Maddison and Maddison 1992, for a full discussion of these issues). For some trees, it is assumed that the clock of this unidirectional, simple, monocausal history is also as regular as clockwork. These are the molecular biologists of the 1980s and 1990s who sought mitochondrial Eve, our shared progenitor, and who attacked phylogenies produced in other disciplines as being historically inaccurate. Their phylogenies were based on the assumption of a regular rate of mutation, so that current percent of difference from the root stock represented the amount of time since divergence. Given the overwhelming evidence for differential rates of mutation, the quest today is to find sites on genes that can serve as relatively reliable timekeepers. So trees as representations of life or knowledge are a problem for the white-coated molecular biologist as well as for the unwashed postmodernist. There is nothing surprising in these convergent representational problems—Gerald Holton (1988), for example, produces a number of others from the fields of history, physics, and mathematics over the past century and a half.

The tree of life maps the diversity of life. To do so, it breaks the web of life into countable units. These units are assumed to be entities in the world, although there is fierce debate over their nature. These countable units are then aligned in a regular (in some cases, metronomic) historical time, in which there is no turning back and no speeding up. The unchanging species is mapped onto the flat time. Although temporality is thus doubly invoked, and is clearly central to the discourse, it is often invisible in discussions of the tree. And yet this folded temporality is precisely our effort to map the world, in all its complexity, onto a linear, featureless time.

A tree, then, is an expression of the modality of particularity. It is an attempt to represent all of life in its infinite diversity within a single representational structure, so that even the ephemeral mayfly can find its place. Such modalities are constitutive of much biodiversity discourse today. Stewart Brand's all-species foundation (http://www.all-species.org/) as well as the All Taxa Biodiversity Inventories (e.g., http://iris.biosci.ohio-state.edu/projects/atbi_db.html) are recent multi-million-dollar efforts to produce better lists of life on this planet. This is the other side of the coin from the modality of implosion described previously. With the modality of particularity, we find background stasis. Events—which would involve entities, a place, and a time—are systematically excluded from the representational framework, thus creating background stasis and an argument for taxonomists about whether cladistic trees have roots (represent change over what we have seen to be a deliberately smoothed, anisotropic time) or are formal devices for assigning names. The result is a packaging of species that guarantees humans some kind of immunity from the flow of natural time (we are a single, well-defined species) and so creates room for a foregrounding of the changes we induce on the external object nature.

The Emergent Globalizing Ethnos

In a modality of implosion, representations are made of several registers within a single structure; the representations are imploded into a singular form rather than exploded into full detail. A rich example of this comes from the Lukasa memory device (Roberts et al. 1996), which contains topographical, historical, property, and political relations within a single, handheld board (figure 5.5). Within biodiversity discourse, the standard modality of implosion is scarcely so rich. This modality seeks to reduce plants, animals, viruses, bacteria, and

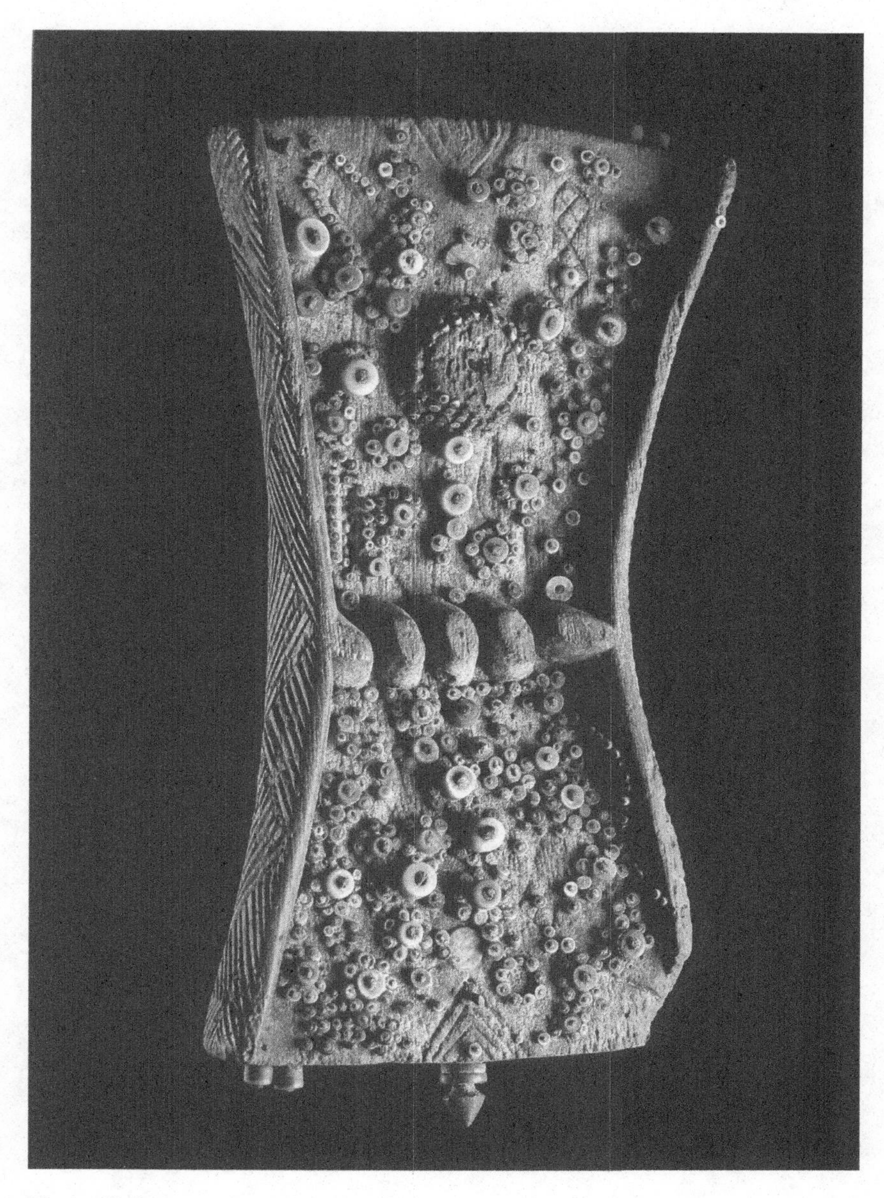

Figure 5.5
Lukasa memory device. *Source:* M. N. Roberts et al., *Memory: Luba Art and the Making of History* (New York: Museum for African Art, 1996).

so forth into a single "biodiversity value," which can used in making policy decisions about what to save and how to save it: it might turn out more efficient in biodiversity terms to let a rare species die out if a sister species, with much the same genetic stock, is unthreatened, for example.

Temporal orientation (how we conceive of the present, past, and future and the flow) is central to the operation of contemporary modalities of implosion and particularity; and this orientation simultaneously operates on the register of the nature of the world and the operation of our political economy. In the case of the coin, we saw the mapping of infinity onto an amount. In that process, which revolved around the construction of a nature/culture divide, we saw the cutting up of emergent forms into units that could circulate within Newtonian space and time. In the case of the tree, we saw the breaking down of complex historical time into regular, calculable units. Common to the enterprise of both modalities is the incorporation of natural objects into cultural discourse. Describing biodiversity and its value through these modalities involves creating databases out of which only certain sets of narrative form can emerge—the story of the house that Jack built, a simple story that proceeds in a regular rhythm.

Attention to these modalities draws us to a (global) anthropological reading of biodiversity discourse. This new discourse is confronting other ways of knowing (referred to as indigenous, local, and vernacular—all terms have their problems if you think about their other—knowledge), not as another myth system revolving around the construction of the nature/culture divide, and yet which has some valuable nuggets of truth, but as a truth system revolving around the way the world is. This seems unfair. Money is not the optimal symbolic form for bringing together the various actants in mutual accord. As currently being worked through, money discourse encourages the evacuation of event-based ontologies through excluding just that sort of memory we should be exploring in order to deal with planetary management. In so doing, it settles the question of the mediation between inside and outside (nature and culture) in a way that is ineluctably ethnocentric. Ethnocentric because the discourse is structured by the way "we" handle the nature/culture divide; it casts the world and time according to the very singular *oikos* of our emergent globalizing late capitalist *ethnos.* Stable tokens beget tokenism.

Keeping Knowledge Local

We are currently, as a globalizing presence, seeking to preserve many kinds of diversity—linguistic, cultural, ethnic, genetic, to name but a few. And yet that preservation has its price; ethnic and linguistic diversity can best be maintained by sequestration, which is good for the connoisseur of the diverse but not necessarily for the diverse themselves. Databases are often seen as a good site for preservation without politics: from the Mayan cultural atlas (Toledo Maya Cultural Council and Toledo Alcaldes Association 1997) through the efflorescence of museums of indigenous knowledge. Really listening to other ways of knowing entails more than databasing. After all, indigenous knowledge tends to end up in text fields in scientific databases: collocated with the real data but unmanipulable and hence unusable. How should we record and remember other ways of knowing?

In May 2001 I attended a three-day workshop on indigenous knowledge. Our workshop was a follow-up to the Innovative Wisdom conference in Florida the previous year, responding to Article 26 of a World Conference of Science report (which led to controversy in the journal *Nature*). The offending article read as follows: "26. That traditional and local knowledge systems as dynamic expressions of perceiving and understanding the world, can make and historically have made, a valuable contribution to science and technology, and there is a need to preserve, protect, research and promote this cultural heritage and empirical knowledge" (ICSU-UNESCO 1999). The International Council of Scientific Unions (ICSU) took collective umbrage at this statement, arguing that there was no such thing as traditional knowledge, and if there was it certainly had not contributed to science and technology, and if it had in the past contributed then there was certainly no need to promote it in the future since the light of modern science could shine brighter and cheaper into local territory. I was part of a group of historians, philosophers, and sociologists of science charged with defending Article 26 to ICSU.

"Innovative wisdom" is a nice phrase. There is a great dearth of words for what it is that locals have and universals do not. If we call it "indigenous knowledge," then we are denying the role of the urban dweller who has moved in from another area, or the *mestizo*, the half-caste. Many anthropologists and others also oppose the term "traditional," which implies knowledge that is lodged in an eternal past (where our knowledge uncovers the eternal present). At a recent conference of signatories to the Convention on Biological Diversity in Nairobi, African delegates said that they felt insulted if their knowledge was referred to as traditional; for them, "indigenous" refers to community and is the correct unit of analysis. In Latin America, "indigenous" is also considered a reasonable epithet. However, delegates from Morocco don't consider themselves indigenes and reject "traditional," which has connotations of being backwater. ILO convention 169 refers to the rights of indigenous and tribal peoples in independent countries; the Nairobi conference ended up adopting "indigenous and local." In the (highly ironic) Museum of Jurassic

Technology in Los Angeles, an exhibit on local knowledge refers to a society established in Edwardian England for the Restitution of Decayed Intelligence! Of course there's always *pensée sauvage,* but that wildflower is untranslatable.

Of course, one is giving the game away at the outset if one says that some knowledge is local—since by extension, there exists some other knowledge that is universal. We produce knowledge that is true in all places at all times; they produce knowledge that is particular to their region. Bruno Latour and Harry Collins have in different ways shown us that our "universal" knowledge is restricted to highly localized space and time: the space and time of the laboratory. When a science test starts with the phrase "All things being equal . . . ," then asks how long a falling body will take to light, it points in the direction of all the work that is done in making other things equal—excluding vibrations, foreign products invisible to the naked eye, weather conditions so that in this very small, highly localized place, the laboratory, universal knowledge can be produced. Mark Meadow and Bruce Robertson's Microcosms project (http://www.microcosms.ihc.ucsb.edu/) shows just what a strange array of objects university knowledge is conjured out of: they inventoried then exhibited collections of objects in libraries and in departments across the University of California system, showing they are alike these are to cabinets of curiosities. Further, as Latour (1987) points out, this highly localized knowledge can only travel along very sparsely populated networks (from one laboratory to another) before it begins to make the necessary set of alliances and affect (or effect, depending on your point of view) the world. So I don't want to say that at universities we produce universal knowledge and that in the outback others produce local knowledge. Knowledge is always firmly tied to a locality.

And to a temporality. ICSU took particular offense at the phrase "traditional and local knowledge" because the state of Kansas had recently voted evolutionary theory to be nonscientific and thus not teachable as science in the classroom. How, they asked, could we maintain distance between church and state—a long-cherished divide, constructed over several centuries in Western science—without being able to challenge hocus-pocus-like the oxymoronic "creationist science" or the knowledge of the shaman or astrologer? Local knowledge became *at this time* something that true science had to challenge; at other times the issue would not have arisen, or fear of political fallout would have suppressed the assertion. The historical junctures at which universalism asserts itself are invariably those where other knowledge systems threaten the monotheistic trick borrowed from that other great universalizing tradition.

Careening through our universal space is the Scylla and Charybdis of local knowledge—the Scylla being that its space cannot be recognized as socially performed and it cannot be historically constructed if it is to be knowledge, and the Charybdis being that if it does not conform to (or overly complicates) traditional rules of intellectual property, then it will be ignored or worked around. So it can only be knowledge at the price of denying its very nature.

So Why Keep Knowledge Local and Local Knowledge?

Well, the first answer is that it is anyway. All knowledge, as pointed out earlier, is irredeemably local. Howard Becker (personal communication) has a nice mantra for the state of scientific knowledge: "They used to believe, but now we know." The astonishing thing—at any level of temporal granularity—is that so many think that our most cherished notions are established for all time. Now we know. It is a terrible hubris to say that one has access to the only way of knowing. We ourselves—whoever that may be—have several, contradictory, very powerful ways of knowing. Attention beyond reason to a single way of knowing is attaching to a fetish-demanding obsession, not treading the one, true, right, and only path. Michel Serres (1980) long ago stated that his role as a philosopher of science was to keep the spaces between the disciplines open. Disciplines did what they did very well, but they also had an imperial tendency to deny any knowledge that did not conform to their way of knowing.

More substantively, I could perhaps call to mind the beautiful lines in Leibnitz et al.'s *Writings on China* (1994) about the difference between Chinese and European knowing. Leibnitz argues that where the West has mathematical and logical knowledge down, the East understands the Art of Society. This resonates with much current Vulgate about the Other—they know how to live; how to laugh; how to subsist in harmony with nature. The power and the glory are on our side, but they get the rest. However, Leibnitz also hails as one of the greatest discoveries of all time the invention of binary notation—which he discovered through the *I Ching* and its hexagrams. The powers of binary arithmetic are not to be denied in the computer age. Crucially, however, the argument can be made that the binary has become a tool for thinking within our culture, as it was in China. One has only to look at the binary oppositions of the structuralists to see the binary's penetration into anthropological and philosophical discourse in the 1980s. If we can say that we have learned a tool for thinking with, rather than wrested a nugget from a blank slate, then we can say that we are listening forth to all knowledge.

Conclusion

What's It All About? (The Practice of Memory and the Memory of Practice)

Time is the hardest thing, as well as being the most evanescent. We have, so it is said, become disciplined to clock time—indeed, the epoch of this book is coeval with that exercise. I certainly cannot imagine any other way of getting through all the things I have to in my day without a pretty well-ordered calendar. We recognize our executives by their particular time disciplines (pagers, cell phones, email reminders, electronic calendars), as constricting as the somewhat inevitable tie is for the busyfull businessman. For those leading a busyless existence, languorous idyll is often still punctuated by the rhythm of *Days of Our Lives* or the obligatory weekly visit to the unemployment office.

In all this, where is the past? It's all around me. There are books on my shelves that serve as trophies of past pleasure rather than, in general, tokens of future remembrance. I'll reread them (some) when I retire (but I don't want to retire). There are the photos in my drawers that someday I'll look at again. Maybe that's a retirement project: putting them in order as I did my parents' to mark my mother's death and attempt to bond with my father—who purified our family past with such passion that little percolated to me. Of course, there are the email traces. I love those—memory is cheap; and so I now have random access to all the emails I've written over the past ten years. Not that I ever actually go back, but I imagine that some day they will prove to be incredibly useful. Anyway, it's comforting to know they're there. And then there are my memories. I grew up in the heady 1960s—full of heads with a revolutionary future just around the corner and the past (or at least short-term memory) going up in a cloud of smoke. I guess I think back to that, certainly not as a golden age; I am leery of golden ages that must have existed (for so I am told) in the proximate past, just out of my reach but apparently available to many of my friends and dear colleagues. When the university was a liberal arts institution. When we used to care. When things were simple. When we were the way we were.

And then there's history. Well, that's my life's work. Not to respond to the interpellation "Zakhor!"—it's not for me about events or singularities (which divide us) in that way. Rather, history is what lets me merge into a more general understanding of how I am part of it all. That I am part of stardust. That my cells are made of (possibly) symbiont entities. That viruses are part of my shared history with plants, protests, and archaea: they grew up with me; indeed, even the cell nucleus itself (the protected core of my genetic identity) is arguably a viral invasion. That my species—to hang a familiar word on an eldritch concept—has coevolved with its environment. That geodiversity has begotten biodiversity—and so that the complexity of the living world stands in intimate relation to the complexity of the inaminate. That the tears in my eye evoke the primal ocean. Bergson ([1939] 1968) has best described the way that memory infuses matter. I just wish I could remember what he said.

And thus the exercise of memory is for me transcendent as well as immanent. It is also political as well as personal. A totalizing state, not unlike mine own, is one that constantly effaces the past as memory to reify it as empty triumph. We have seen two aspects of this logic. First was in the Monte Cristo story told in the beginning: the prison does not, as Dumas points out, need to remember the crimes and individuality of the Count; just that he has a number and that they have a series of regulations for people with numbers. Comte gave us the second: if we have the truth now (and that truth is complex) then we really don't have time or energy to follow through past follies in order to attain the timeless reality of science. We can cut to the chase. Bruno Latour (1996b) has in recent years begun to explore the "police" action of science: one dimension of that police action is the effacing of the multiple past to replace it with the singular present converging ever faster on the end of history.

So you ask "Why memory practices?" and I respond "Well, they are immanent and transcendent, and I guess personal and political as well." This is the kind of portmanteau response I know and love from my structuralist forebears (and that drove René Girard (1972, 336) crazy). After all, Lévi-Strauss concludes at the end of his four-volume mythologies that maybe it's about the nature of reality or maybe about the nature of mind—or then again maybe about the organization of society (Lévi-Strauss 1971, introduction). Let me start the trek back to the specific set of memory practices we have looked at in this book.

What fascinated me from the start of this project, lo these many years, was the resonance between what Lyell said about the earth and what he said about the discipline of geology. I have been worrying at that for years. A catastrophic past for both was wrong; a uniform past for both was right. The establishment (proof) of the one resonated with the establishment (foundation) of the other. With a uniform past, the memory problem went away, in just the same way as it had for Comte and for the totalizing state. You no longer had to remember things past, just the names of the processes and the rules that made them balance. The rules made the present of the earth as timeless (and without vestige of origin or sign of end) as the present of the pursuit of geological science. Or, to be quite precise—and I thank Katie Vann for discussions leading to this—to the proximal future of the discipline: that ideal present which is always just around the corner and never quite there.

Wherever I turned—paradoxically in an age of progress writ large (though we live an age that has systematically evacuated progress of all positive features, leaving it to gnaw on its etymological entrails, as onward march)—I found over the past two hundred years the traces of an ideal eternal present. An end to history. Central to this eternal present were different modalities of suppressing the secular past. In the 1830s, as we have seen, geology turned secular history into balanced bookkeeping, turned a series of events in particular places into a series of rules about the balance between sea and land. In the 1950s, cyberneticians created resonance across systems of all kinds and at all scales at the price of particular memory. It could not matter that humans had memories, because we were effectively Turing machines. It could not matter that cells had particular histories rather unlike those of seals and deals, since the organization of each resonated with the others. In the current epoch, the earth with a turbulent past and uncertain future is constantly cast in terms of "preservation" or "preventing climate change." The latter is with a view to spiriting the effects of humans out of the story of the history of the earth (what we do should, in Lyell's terms, operate a moral and not a physical effect).

Not that I am in favor of belching smokestacks and the roar of internal combustion engines: all I am doing here is questioning the tropes that are used to describe them. They are tropes that operate a great divide between our knowledge, which is timeless and without a past trajectory (it is always already there) and other folks' knowing, which is tied oh-so-closely to their environment and their traditions. We need to preserve their memories and their diversity, precisely because we are the ethnos without memory: particularity is always already other. Their knowing is useless precisely because it is tied to its own roots.

I have discussed throughout this book how memory itself gets configured very differently in different information infrastructures. There is a trite dimension to this remark: after all, no one is really going around saying that ledger books are like file folders are like object-oriented databases. Where it gets some purchase, however, is through its reach: how materially and metaphorically it marks a movement between insides and outsides. Materially first. I was most struck by Derrida's comment that our consciousness is different now because of the new information technology we have—in particular, through the mediation of the new memory prostheses. I am fundamentally marked by my information technology: it's not that an unchanging identity is expressed differently because of the new recording media (photographs, recorders, databases); but that a new identity is formed through its imbrication in those very media.

Now metaphorically. Take the analogy of one of the great early memory devices in computing: the Williams tube. The tube was a few feet in length and filled with mercury, which may have been quicksilver but furnished in this case slow memory. An intermediate calculation could be consigned to a pulse that would travel very slowly along the tube (relatively) and then be delivered back to the calculator when other operations had prepared it to process the result. Memory was distributed across media heterochronously so as to produce a result (a present) in which the past would have been simple, linear, and sequential (a set of additions and subtractions leading to a total). Our own identities work that way: we loop memory out heterochronously across a range of media and materials (friends, conferences, photos, letters, date books . . .) so that at any one time we can, in theory, draw it all together along the single timeline of our past. Indeed, as we learn to become kinds through confessional shows or documentaries, our stories weave together with those of others. We use information infrastructure, pace Lacan, to speak us. Resonant with this material distribution of our memory traces is their metaphorical representation as we describe ourselves. My life is an open book. I'll file that away for future reference. Those evangelicals program their followers to say the darnedest things. You need to give me your feedback on this stuff. She has a photographic memory. Store that one away. And so forth. We use information infrastructure to speak with.

A Trip Down Memory Lane (The Long Haul)

The period covered in this work is essentially the period of the growth of global planetary management (Serres 1990; Elichirigoity 1999): our "natural contract" and our "social contract" each demand the same effort of information integration in order to develop. We now have global readings of insolation (the amount of energy coming in from the sun), the amount of that energy that is converted by living organisms (through the surrogate measure of carbon fixing) and the percentage that is consumed by people. People have become the obligatory passage points bar none in the political economy of the planet: we take a dominant percentage of incoming solar energy; we control over 95 percent of free-flowing water resources; we have since our inception been carrying out a process of sustained extinction of other species that is culminating in a catastrophe of geological proportions. We take similar control over each other: disciplining ourselves to temporalities and ideologies that allow us to be governable.

How do we gain such empire over the present? Information integration has a lot to do with it—the ability to collect data from numerous disparate resources, collocate it (through the production of (im)mutable mobiles), and then use it to plan the future. In order to achieve this integration, we weave together stories about the past of the earth, the past of the cosmos, and the past of our knowledge out of a tangle of threads. Our reading of the past is generically under the description of the present set of entities and phenomena; this was Lyell's central methodological breakthrough, and it has stayed with us. Past time is the same as present time; past entities of the same order as present entities. And the present is effectively perfect, for Lyell through to most biodiversity science today. The goal is to stop the extinctions, stop climate change, stop up the hole in the ozone layer. Ecosystems and people should, within this frozen present, maximize productivity. The only good ecosystem is a productive ecosystem—giving us goods and services just like the Third World does (and indeed it is remarkable how moves to preserve linguistic, ethnic, and cultural diversity express the need to freeze the present within a model of global productivity). The great trick, with respect to the past, has been to project present entities and processes back into the past—leaving the present as the natural and timeless outcome of a teleological process. The memory practices that we have examined are globally about producing this frozen, perfect, productive present as integrally spiritual epiphany and political reality.

I have sketched some ways in which the traffic between the understanding of the world and the empire of the West has become imbricated. Most notable here is the shadowy role of bureaucracy: the tools of bureaucracy (the list and the surrogate coin as the degree zero of information) act as a metaphor (and literally as *metaphoros* or "transport") enabling the creation of a universal yet entirely ethnocentric past. For geology, we saw the metaphor of the archives weaving together the past of the earth and the operation of nascent governmentality. For cybernetics, the metaphor of the control revolution (with the feedback loop tied to the governor in the steam engine being the guarantor of repetition) brings together all possible histories and meditates on the political economy of control. For biodiversity, the metaphors of the bank and the database allow us to read life as information and simultaneously to meditate on the economy of our globalizing empire.

We have a restricted set of stories that we tell and can tell about the past. Tort (1989) demonstrated so brilliantly for classification systems the propagation of genetic classification through many social and natural spheres in the nineteenth century: a good bureaucratic trick travels well. Similarly, the effectiveness of universal timelines and isotropic time and space have been demonstrated through a set of well-traveled bureaucratic developments—from the mundane file folder organized by class and date to the development of object-oriented databases (themselves, of course, subject to the remorseless ticking of the internal computer clock). Out of these tools of empire we create a past of a very particular sort—one in which there is really but one line into the "Mnemonick Deep." I have argued for a deeper consideration of the role of our memory practices as the site where ideology and knowledge fuse. My preference would be for a harlequin's coat (Serres 1991; cf. Turnbull 2000) of a past, where contradictory temporalities and entities could be explored. We need to hold the past open so that we do not hypostasize and freeze the present, and by extension limit our own future.

This has largely been a book about the *longue durée*. I have concentrated centrally on what has happened over the past few hundred years as the database has developed as a central symbolic form. Along the way, I have not felt the need to operate one divide so characteristic of our current era—between modernism and postmodernism. This, for me, is in general just one of a series of breaks that we operate with the past in order not to have to go back to it—just as Lavoisier took out the previous century's language of chemistry, or eleventh-century France offered the appearance of a radical break with the past as a way of shielding continuities (Geary 1994). The past is a heavy weight, and we have continuously since the beginning of the nineteenth century liked to describe our current epoch as radically different from all before it—the acceleration of information acquisition and storage requiring perhaps a more frequent spring cleaning of the attic than it had for earlier generations.

If I had to tell a single story out of the many that have gone through this work, it would be this. The archive, we saw at the beginning, is both sequential and jussive—containing both the law of what can be said and the moment from which it is first uttered. So Lyell gives us an archive for geology, centered on the metaphor of the archival commissioner. From this moment on, the practice of the science of geology will mirror the true past of the history of the earth—the two will be mapped onto an ideal lawlike second nature. Comte came closest to describing this process for science in general; when there is too much history, only rational reconstruction will do the job of holding the past in a memorable form. Cyberneticians too inaugurated an age in which a new kind of archive could be kept, this time a past that could be generated from first principles across a range of media. Just as a single computer program could express the reproduction of potholes in a road, sheep in a pasture and ideas in a field. Just as the discipline of geology could be synchronized with the history of the earth, so could the interdiscipline of cybernetics (through the magic of recapitulation) with the history of life. The origin of the earth, for Lyell, was thrown into deepest past; as was the origin of information processing for cyberneticians (who put it in the first quantal phase change). Pushing origins into the inaccessible past was a trick that allowed them to evacuate the more recent past of specific content, and so to operate a foundational act in the present synchronizing the human and natural worlds. The resonance between matter and metaphor mediated by information technology emerged, in the process, as a powerful subterranean tool.

Which brings us to biodiversity databases today. Now that the Savings and Clone company is offering to bank our pets' genes and clone them at will (cats now, dogs just down the road), we can begin to get a better grasp of what was being inaugurated in the archive whose features we have described in this book. Our ethnos without a past now spread across the space of the globe has, over the past few hundred years, done a fairly good job of synchronizing first and second nature, on the one hand, and matter and metaphor, on the other. Information storage and retrieval has been central to this work. Sequentially. From this moment on (the end of the eighteenth century), we (the globalizing "we") began the task of somewhat systematically taking an inventory of life (the list, beginning with Linnaeus, who first gave us names). We also began the massive bureaucratic task of commodifying life, by way of the coin (beginning with Malthus, who gave us the necessity of planetary management). We see life through the prism of the coin and the list, the program, and the database. Jussively. Because we package the past into these bundles, we can only distinguish stories that operate within these tightly coupled registers.

Not that there's anything wrong with that. If it is ineluctable, as it appears to me now, then at least we should be modest about our stories. Science as we know it and globalization go very well together; that's just the way we deal with the nature/culture divide. But there is a more positive message than this. If our ways of knowing are invariably about our ways of living; then social and political experimentation can be recognized as integrally social, ontological, and epistemological. We are what we eat, and we think what we practice.

Peroration (Of Passing Interest)

Memory practices matter because they are what carries the past along with us into the future: they are what makes our current reality true and our future—

in will if not in deed—controllable. They matter for us (the globalizing ethnos) because our tricks of the trade for evacuating the past of specificity through transmuting ourselves and the world into kinds are what allow us to perpetuate the myth that we are the measure of all things.

Just because the past is over doesn't mean that there is a truth about "wie es eigentlich gewesen ist." The work of creating partial objects and conjuring them into a given, small set of trajectories is a work in the present of expanding our empire and our knowledge. If we want the future to be other than it seems to be turning out, we must create a past that is other than it seems to have turned out. People, planets, and purgatory (Le Goff 1984) deserve multiple pasts. Only an open past can unlock the present and free the future.

References

Abbott, A. 1988. *The System of Professions: An Essay on the Division of Expert Labor.* Chicago: University of Chicago Press.

Adam, B. 1990. *Time and Social Theory.* Cambridge: Polity Press, in association with Blackwell Publishers.

Adams, D., and J. Lloyd. 1990. *The Deeper Meaning of Life.* New York: Harmony Books.

Agamben, G. 1993. *Infancy and History: The Destruction of Experience.* London and New York: Verso.

Ager, D. V. 1993. *The New Catastrophism: The Importance of the Rare Event in Geological History.* Cambridge: Cambridge University Press.

Ainsworth, W. 1964. "Electrolytic Growth of Silver Dendrites." *Science* 146: 1294–1295.

Allègre, C. J. 1992. *From Stone to Star: A View of Modern Geology.* Cambridge, MA: Harvard University Press.

Althusser, L. 1996. *Lire le capital.* Paris: Presses Universitaires de France.

Anderson, W. R. 1991. "Should We Change the Code? Concerns of a Working Taxonomist?" In *Improving the Stability of Names: Needs and Options: Proceedings of an International Symposium, Kew, 20–23 February 1991*, edited by D. L. Hawksworth, 95–103. Königstein/Taunus, Germany: Koeltz Scientific. Published for the International Association for Plant Taxonomy in conjunction with the International Union of Biological Sciences and the Systematics Association.

Anonymous. 1836a. *Monthly Notices of the Astronomical Society of London, Containing Abstracts of Papers and Reports of Proceedings* 3.

Anonymous. 1836b. "Life and Works of Baron Cuvier." *Edinburgh Review* 62: 265–297.

Apter, M. J. 1975. "Cybernetics: A Case Study of a Scientific Subject-Complex." *Keele Sociological Review Monograph: The Sociology of the Sciences:* 93–115.

Arbib, M. 1989. *The Metaphorical Brain 2: Neural Networks and Beyond.* New York: Wiley.

Arnot, N. 1827. *Elements of Physics: or Natural Philosophy, General and Medical, Explained Independently of Technical Mathematics.* London: J. Murray.

Ashby, W. R. 1956. *Introduction to Cybernetics.* London: Chapman and Hall.

Ashby, W. R. 1958. "General Systems Theory as a New Discipline." *General Systems Yearbook* 3: 1–6.

Ashby, W. R. 1962. "The Self-Reproducing System." In *Aspects of the Theory of Artificial Intelligence: The Proceedings of the First International Symposium on Biosimulation, Locarno, June 29–July 5, 1960*, edited by C. A. Muses, 521–536. New York: Plenum.

Ashby, W. R. 1978. *Design for a Brain: The Origin of Adaptive Behaviour.* London: Chapman and Hall.

Auger, P. 1960. "Introduction." In *Proceedings of the Second Interational Association for Cybernetics, Namur, 1958*, i–xii. Paris: Gauthier Villars.

Avi Silberschatz, M. S., and Jeff Ullman. 1994. "Database Systems: Achievements and Opportunities." In *Readings in Database Systems*, edited by M. Stonebraker, 921–931. San Francisco: Morgan Kaufmann.

Babbage, C. 1830–1840. "Correspondance." *British Museum Add. Mss. 47545–47547.* London.

Babbage, C. 1832. *On the Economy of Machinery and Manufactures.* Philadelphia, PA: Carey Lea.

Babbage, C. 1837. *The Ninth Bridgewater Treatise: A Fragment.* London: J. Murray.

Barrowclough, G. F. 1992. "Systematics, Biodiversity and Conservation Biology." In *Systematics, Ecology, and the Biodiversity Crisis*, edited by N. Eldredge, 121–143. New York: Columbia University Press.

Bateson, G. 1967. "Cybernetic Explanation." *American Behavioral Scientist* 10(8): 29–32.

Baudrillard, J. 1995. *The Gulf War Did Not Take Place.* Bloomington: Indiana University Press.

Beaumont, L. E. de. 1832–1843. "Collège de France, 1832 à 1843." In *Papiers, Boîte 9.* Paris: Institut de France.

Beer, S. 1959. *Cybernetics and Management.* London: The English Universities Press.

Beer, S. 1975. *Platform for Change.* London: Wiley.

Bell, R. E., M. Studinger, A. Tikku, G. Clarke, M. Gutner, and C. Meertens. 2002. "Origin and Fate of Lake Vosok Water Frozen the Base of the East Antarctic Ice Sheet." *Nature* 416: 307–310.

Beniger, J. R. 1986. *The Control Revolution: Technological and Economic Origins of the Information Society.* Cambridge, MA: Harvard University Press.

Benjamin, W., and H. Arendt 1986. *Illuminations.* New York: Schocken Books.

Benner, J. 1835. *Commentaire philosophique et politique sur l'histoire et les révolutions de France, de 1789 à 1830.* Paris: Treuttel et Wurtz.

Bensaude-Vincent, B. 1989. "Lavoisier: Une révolution scientifique." In *Eléments d'histoire des sciences,* edited by M. Serres, 363–386. Paris: Bordas.

Berg, M. 1997. *Rationalizing Medical Work—Decision Support Techniques and Medical Problems.* Cambridge, MA: MIT Press.

Berger, J., and J. Mohr 1975. *A Seventh Man: A Book of Images and Words about the Experience of Migrant Workers in Europe.* Harmondsworth and Baltimore: Penguin.

Bergery, C-L. 1829. *Economie industrielle, t.1: Economie de l'ouvrier.* Metz: Thiel.

Bergson, H. [1939] 1968. *Matière et mémoire.* Paris: Presses Universitaires de France.

Bisby, F. A. 1988. "Communications in Taxonomy." In *Prospects in Systematics,* edited by D. L. Hawksworth, 277–291. Oxford: Clarendon Press and Oxford University Press. Published for the Systematics Association.

Bohr, N. 1960. "Quantum Physics and Biology." In *Symposia of the Society for Experimental Biology* 12: 138–163. Cambridge: Cambridge University Press.

Borges, J. L. 1964. "The Fearful Sphere of Pascal." In *Labyrinths: Selected Stories and Other Writings,* 189–192. New York: New Directions.

Borges, J. L. 1998. *Collected Fictions.* New York: Penguin Books.

Boulter, M. C., et al. 1991. "The IOP Plant Fossil Record: Are fossil plants a special case?" In *Improving the Stability of Names: Needs and Options: Proceedings of an International Symposium, Kew, 20–23 February 1991,* edited by D. L. Hawksworth, 231–242. Königstein/Taunus, Germany: Koeltz Scientific. Published for the International Association for Plant Taxonomy in conjunction with the International Union of Biological Sciences and the Systematics Association.

Bowker, G. C. 1994. "Information Mythology: The World of/as Information." In *Information Acumen: The Understanding and Use of Knowledge in Modern Business,* edited by L. Bud-Frierman, 231–247. London: Routledge.

Bowker, G. C. 1994. *Science on the Run: Information Management and Industrial Geophysics at Schlumberger, 1920–1940.* Cambridge, MA: MIT Press.

Bowker, G. C. 1994. *Science on the Run: Information Management and Industrial Science at Schlumberger, 1920–1940.* Cambridge, MA: MIT Press.

Bowker, G. C. 1997. "Lest We Remember: Organizational Forgetting and the Production of Knowledge." *Accounting, Management and Information Technology* 73: 113–138.

Bowker, G. C., and S. L. Star 1999. *Sorting Things Out: Classification and Its Consequences.* Cambridge, MA: MIT Press.

Bowser, C. J. 1986. "Historic Data Sets: Lessons from the Past, Lessons for the Future." In *Research Data Management in the Ecological Sciences*, edited by W. T. Michener, 155–179. Columbia: University of South Carolina Press.

Boyle, T. J. B., and J. M. Lenne 1997. "Defining and Meeting Needs for Information: Agriculture and Forestry Perspective." In *Biodiversity Information: Needs and Options: Proceedings of the 1996 International Workshop on Biodiversity Information*, edited by David L. Hawksworth, 31–353. Oxon, UK, and New York: CAB International.

Brandenburg, W. A. 1991. "The Need for Stabilized Plant Names in Agriculture and Horticulture." In *Improving the Stability of Names: Needs and Options: Proceedings of an International Symposium, Kew, 20–23 February 1991*, edited by D. L. Hawksworth, 23–31. Königstein/Taunus, Germany: Koeltz Scientific. Published for the International Association for Plant Taxonomy in conjunction with the International Union of Biological Sciences and the Systematics Association.

Braudel, F. 1973. *Capitalism and Material Life, 1400–1800.* London: Weidenfeld and Nicolson.

Brenchley, P. J., and D. A. T. Harper. 1998. *Palaeoecology: Ecosystems, Environments, and Evolution.* London: Chapman and Hall.

Briquet, J. 1906. *Règles internationales de la nomenclature botanique adoptées par le Congrès International de Botanique de Vienne 1905 et publiées au nom de la Commission de Rédaction du Congrès.* Jena: Verlag von Gustav Fischer.

Briquet, J. 1934. *International Rules of Botanical Nomenclature Adopted by the Fifth International Botanical Congress, Cambridge, 1930.* London: Taylor and Francis.

British Sessional Papers. 1818. London: HMSO.

Brooks, F. T., and T. F. Chipp, eds. 1931. *Fifth International Botanical Congress, Cambridge, 16–23 August 1930, Report of Proceedings edited for the Executive committee by F. T. Brooks and T. F. Chipp.* Cambridge: Cambridge University Press.

Brün, G. F., and M. 1999. "Natural Environment and Human Culture: Defining Terms and Understanding Worldviews." *Journal of Environmental Quality* 28: 1–10.

Buchez, P-J-B. 1833. *Introduction à la science de l'histoire; ou science du développement de l'humanité.* Paris: Paulin.

Buckland, M. K. 1997. "What is a 'Document'?" *Journal of the American Society for Information Science* 48(9): 804–809.

Buckland, W. 1836. *Geology and Mineralogy Considered with Reference to Natural Theology.* London: W. Pickering.

Buitenen, J. A. B. van. 1973. *The Mahabharata.* Chicago: University of Chicago Press.

Bulletin de la Société Géologique de France. 1833. Paris.

Burchfield, J. D. 1990. *Lord Kelvin and the Age of the Earth.* Chicago: University of Chicago Press.

Burnett, D. G. 2003. "Mapping Time: Chronometry on Top of the World." *Daedalus* (Spring): 5–19.

Bush, V. 1945. "As We May Think." *Atlantic Monthly* 176(1): 101–108.

Butler, S. 1872. *Erewhon, or, Over the Range.* London: Trèubner.

Cain, A. J. 1958. "Logic and Memory in Linnaeus's System of Taxonomy." *Proceedings of the Linnean Society of London* 169: 144–163.

Callon, M. 1994. "Is Science a Public Good?" *Science, Technology and Human Values* 19(4): 395–424.

Campbell-Kelly, M. 1994. "The Railway Clearing House and Victorian Data Processing." In *Information Acumen: The Understanding and Use of Knowledge in Modern Business,* edited by L. Bud-Frierman, 51–74. London: Routledge.

Candolle, A. de. 1867. *Lois de la nomenclature botanique.* Paris: V. Masson et fils.

Cannadine, D. 1992. "The Context, Performance and Meaning of Ritual: The British Monarchy and the 'Invention of Tradition', c. 1820–1977." In *The Invention of Tradition,* edited by E. J. Hobsbawm and T. O. Ranger, 101–164. Cambridge and New York: Cambridge University Press.

Cariani, P. 1993. "To Evolve an Ear: Epistemological Implications of Gordon Pask's Electrochemical Devices." *Systems Research* 10(3): 10–33.

Carpenter, J. R. 1987. Cladistics of Cladistics. *Cladistics* 3: 363–375.

Castells, M. 1996. *The Rise of the Network Society.* Cambridge, MA: Blackwell Publishers.

Chandler, A. D. 1977. *The Visible Hand: The Managerial Revolution in American Business.* Cambridge, MA: Belknap Press.

Chauvet, M. 1991. "The Needs for Stability in Names for Germplasm Conservation." In *Improving the Stability of Names: Needs and Options: Proceedings of an International Symposium, Kew, 20–23 February 1991,* edited by D. L. Hawksworth, 33–38. Königstein/Taunus, Germany: Koeltz Scientific. Published for the International Association for Plant Taxonomy in conjunction with the International Union of Biological Sciences and the Systematics Association.

Chrisman, N., and F. Harvey. 1998. "Boundary Objects and the Social Construction of GIS Technology." *Environment and Planning* A 30: 1683–1694.

Clanchy, M. T. 1979. *From Memory to Written Record, England 1066–1307.* Cambridge, MA: Harvard University Press.

Clanchy, M. T. 1993. *From Memory to Written Record, England 1066–1307.* Oxford and Cambridge, MA: Blackwell Publishers.

Claridge, M. F. 1988. "Species Concepts and Speciation in Parasites." In *Prospects in Systematics,* edited by D. L. Hawksworth, 92–111. Oxford: Clarendon Press and Oxford University Press. Published for the Systematics Association.

Coleridge, S. T. n.d. "Notebooks." *British Museum Add. Mss. 47546*. London.

Collins, H. M. 1985. "Changing Order: Replication and Induction in Scientific Practice." London: Sage.

Colwell, R. R. 1997. "Microbial Biodiversity and Biotechnology." *Biodiversity II: Understanding and Protecting Our Biological Resources,* edited by M. L. Reaka-Kudla, D. E. Wilson, and E. Wilson, 279–287. Washinton, DC: Joseph Henry Press.

Comte, A. [1830–1845] 1975. *Philosophie première; Cours de philosophie positive, Leçons 1 à 45*. Paris: Hermann.

Cox, C. B. 1998. "From Generalized Tracks to Ocean Basins—How Useful Is Panbiogeography?" *Journal of Biogeography* 25: 813–828.

Crane, D. 1972. *Invisible Colleges; Diffusion of Knowledge in Scientific Communities.* Chicago: University of Chicago Press.

Cranston, P. S., and C. J. Humphries 1988. "Cladistics and Computers: A Chironomid Conundrum?" *Cladistics* 4: 77–92.

Cressey, D. 1994. "National Memory in Early Modern England." In *Commemorations: The Politics of National Identity*, edited by J. R. Gillis, 61–73. Princeton, NJ: Princeton University Press.

Cronon, W. 1991. *Nature's Metropolis: Chicago and the Great West.* New York, W. W. Norton.

Cronquist, A. 1991. "Do We Know What We Are Doing?" In *Improving the Stability of Names: Needs and Options: Proceedings of an International Symposium, Kew, 20–23 February 1991*, edited by D. L. Hawksworth, 301–311. Königstein/Taunus, Germany: Koeltz Scientific. Published for the International Association for Plant Taxonomy in conjunction with the International Union of Biological Sciences and the Systematics Association.

Culler, J. 1983. *On Deconstruction.* London: Routledge and Kegan Paul.

Daily, G. C. 1997. "Introduction: What Are Ecosystem Services?" *Nature's Services: Societal Dependence on Natural Ecosystems,* edited by G. C. Daily, 1–10. Washington, DC: Island Press.

Daily, G. C., et al. 1997. "Ecosystem Services Supplied by Soil." In *Nature's Services: Societal Dependence on Natural Ecosystems,* edited by G. C. Daily, 113–132. Washington, DC: Island Press.

Dallenbach, L. 1989. *The Mirror in the Text.* Cambridge: Polity Press.

Daurio, C-P. 1838. *Recherches sur les causes physiques de nos sept sensations, et erreurs des physiciens sur le son et la lumière*. Paris: Desessart.

David, P. 1985. "Computer and Dynamo: The Modern Productivity Paradox in a Not-Too-Distant Mirror." Center for Economic Policy Research, Stanford University, Palo Alto, CA.

Dechert, C. R., ed. 1966. *The Social Importance of Cybernetics.* New York: Simon and Schuster.

Deleuze, G. 1996. *Proust et les signes.* Paris: Presses Universitaires de France.

Dempsey, L., and R. Heery 1998. "Metadata: A Current View of Practice and Issues." *The Journal of Documentation* 54(2): 145–172.

Derrida, J. 1980. *La carte postale: De Socrate à Freud et au-delà.* Paris: Flammarion.

Derrida, J. 1995. *Mal d'archive: Une impression freudienne.* Paris: Galilée.

Derrida, J. 1996. *Archive Fever: A Freudian Impression.* Chicago: University of Chicago Press.

Desrosières, A. 1993. *La politique des grands nombres: Histoire de la raison statistique.* Paris: Éditions La Découverte.

Desrosières, A., and L. Thévenot 1988. *Les catégories socio-professionnelles.* Paris: Éditions La Découverte.

Détienne, M. 2003. *Comment être autochtone: Du pur Athénian au Français enraciné.* Paris: Seuil.

Doré, M. 1830. *De la nécessité et des moyens d'ouvrir de nouvelles carrières pour le placement des élèves de l'Ecole polytechnique, et de l'utilité de créer de nouvelles chaires dans cet établissement.* Paris: Bachelier.

Douglas, M. 1986. *How Institutions Think.* Syracuse: Syracuse University Press.

Dumont, H. J. 1983. "Biogeography of Rotifers." *Hydrobiologia* 104: 19–30.

Dunoyer, C. 1837. *Nouveau traité d'economie sociale, ou simple exposition des causes sous l'influence. desquelles les hommes parviennent a user de leurs forces avec le plus de liberté, c'est-à-dire avec le plus de facilité et de puissance.* Paris: H. Fournier.

Echo du Monde Savant; journal analytique. 1834–1840. Paris.

Edwards, P. N. 1998. "Y2K: Millennial Reflections on Computers as Infrastructure." *History and Technology* 15: 7–29.

Edwards, P. N. 1999. "Data-laden Models, Model-Filtered Data: Uncertainty and Politics in Global Climate Science." *Science as Culture* 8(4): 437–472.

Edwards, T. C., et al. 1995. *Utah GAP Analysis: An Environmental Information System.* Logan, UT: National Biological Service, Utah Cooperative Fish and Wildlife Research Unit, Utah State University.

Eisenstein, E. L. 1979. *The Printing Press as an Agent of Change: Communications and Cultural Transformations in Early Modern Europe.* Cambridge: Cambridge University Press.

Eldredge, N. 1992. "Where the Twain Meet: Causal Intersections between the Genealogical and Ecological Realms." In *Systematics, Ecology, and the Biodiversity Crisis,* edited by N. Eldredge, 59–76. New York: Columbia University Press.

Eldredge, N. 1995. *Dominion: Can Nature and Culture Co-exist?* New York: Holt.

Eliade, M. 1969. *Le mythe de l'âeternel retour; archâetypes et râepâetition.* Paris: Gallimard.

Elichirigoity, F. 1999. *Planet Management: Limits to Growth, Computer Simulations, and the Emergence of Global Spaces.* Evanston, IL: Northwestern University Press.

Engestrom, Y. 1990. "Organizational Forgetting: An Activity-Theoretical Perspective." In *Learning, Working and Imagining: Twelve Studies in Activity Theory*, edited by Y. Engestrom, 196–226. Jyvaskylassa and Painettu: KirjapainoOma Kyssa.

Enserink, M. 1998. "Opponents Criticize Iceland's Database." *Science* 282 (October 30): 859.

Erwin, T. L. 1997. "Biodiversity at Its Utmost: Tropical Forest Beetles." In *Biodiversity II: Understanding and Protecting Our Biological Resources*, edited by M. L. Reaka-Kudla, D. E. Wilson, and E. Wilson, 27–40. Washington, DC: Joseph Henry Press.

Fagot-Largeault, A. 1989. *Causes de la mort: Histoire naturelle et facteurs de risque.* Paris: Librairie Philosophique J. Vrin.

Faith, D. P. 1994. "Genetic Diversity and Taxonomic Priorities for Conservation." *Biological Conservation* 68: 69–74.

Farris, J. S. 1986. "The Pattern of Cladistics." *Cladistics* 1: 190–201.

Felsenstein, J. 1988. "The Detection of Phylogeny." In *Prospects in Systematics*, edited by D. L. Hawksworth, 116–128. Oxford: Clarendon Press and Oxford University Press. Published for the Systematics Association.

Fentress, J., and C. Wickham 1992. *Social Memory: New Perspectives on the Past.* Oxford: Blackwell Publishers.

Fogel, L. J., A. J. Owens, and M. J. Walsh. 1965. "Artificial Intelligence through a Simulation of Evolution." In *Biophysics and Cybernetic Systems; Proceedings of the Second Cybernetic Sciences Symposium*, edited by M. Myles, A. Callahan, and L. J. Fogel, 93–108. Washington, DC: Spartan Books.

Forester, T. 1980. *The Microelectronics Revolution: The Complete Guide to the New Technology and Its Impact on Society.* Oxford: Blackwell Publishers.

Foucault, M. 1975. *Surveiller et punir: Naissance de la prison.* Paris: Gallimard.

Foucault, M. 1982. *The Archaeology of Knowledge.* New York: Pantheon.

Foucault, M. 1991. "Governmentality." In *The Foucault Effect: Studies in Governmentality*, edited by G. Burchill, C. Gordon, and P. Miller, 87–104. Chicago: University of Chicago Press.

Galtier, J. 1986. "Taxonomic Problems Due to Preservation: Comparing Compression and Permineralized Taxa." In *Systematic and Taxonomic Approaches in Palaeobotany*, edited by R. A. Spicer and B. A. Thomas, 1–16. The Systematics Association Special Volume No. 31. Oxford: Clarendon Press.

Geary, P. J. 1994. *Phantoms of Remembrance: Memory and Oblivion at the End of the First Millennium.* Princeton, NJ: Princeton University Press.

Geddes, Patrick, F. R. S. E. 1880–1882. "On the Classification of Statistics and Its Results." In *Proceedings of the Royal Society of Edinburgh* 11: 295–322.

George, F. H. 1965. *Cybernetics and Biology.* London: Oliver and Boyd.

Gerovitch, S. 2002. *From Newspeak to Cyberspeak: A History of Soviet Cybernetics.* Cambridge, MA: MIT Press.

Gillis, J. R. 1994. *Commemorations: The Politics of National Identity.* Princeton, NJ: Princeton University Press.

Gioscia, V. (1974). *TimeForms beyond Yesterday and Tomorrow.* New York, Gordon and Breach.

Girard, R. 1972. *La violence et la sacré.* Paris: Seuil.

Goodwin, C. 1994. "Professional Vision." *American Anthropologist* 96(6): 606–633.

Goody, J. 1986. *The Logic of Writing and the Organization of Society.* Cambridge and New York: Cambridge University Press.

Gould, S. J. 1989. *Wonderful Life: The Burgess Shale and the Nature of History.* New York: W. W. Norton.

Goulder, L. H., and D. Kennedy. 1997. "Valuing Ecosystem Services: Philosophical Bases and Empirical Methods." In *Nature's Services: Societal Dependence on Natural Ecosystems,* edited by G. C. Daily, 23–47. Washington, DC: Island Press.

Graham, L. R. 1972. *Science and Philosophy in the Soviet Union.* New York: Knopf.

Grandsagne, J. B. F. A. de, and V. Parisot 1836. *Philosophie des Science.* Paris: Bibliothèque Populaire.

Gray, B. 1980. "Popper and the 7th Approximation: The Problem of Taxonomy." *Dialectica* 34(2): 129–153.

Gregory, J. 2000. "Sorcerer's Apprentice: Creating the Electronic Health Record, Reinventing Medical Records and Patient Care," xvi, 707 leaves. Ph.D. thesis, University of California at San Diego.

Grehan, J. H. 1994. "The Beginning and End of Dispersal: The Representation of 'Panbiogeography.'" *Journal of Biogeography* 21: 451–462.

Grmek, M. 1990. *History of AIDS: Emergence and Origin of a Modern Pandemic.* Princeton, NJ: Princeton University Press.

Group, T. A. P. 1998. "An Ordinal Classification for the Families of Flowering Plants." *Annals of the Missouri Botanical Garden* 85(4): 531–553.

Gunn, C. R., et al. 1991. "Agricultural Perspective on Stabilizing Scientific Names of Spermatophytes." In *Improving the Stability of Names: Needs and Options: Proceedings of an*

International Symposium, Kew, 20–23 February 1991, edited by D. L. Hawksworth, 13–21. Königstein/Taunus, Germany: Koeltz Scientific. Published for the International Association for Plant Taxonomy in conjunction with the International Union of Biological Sciences and the Systematics Association.

Gyllenhaal, C., et al. 1993. "NAPRALERT: Problems and Achievements in the Field of Natural Products." In *Designs for Global Plant Species Information Systems*, edited by F. A. Bisby, G. F. Russell, and R. J. Pankhurst, 20–37. Oxford: Clarendon Press and Oxford University Press. Published for the Systematics Association.

Hacking, I. 1990. *The Taming of Chance.* Cambridge: Cambridge University Press.

Hacking, I. 1992. "World Making by Kind Making: Child Abuse for Example." In *How Classification Works: Nelson Goodman among the Social Sciences*, edited by M. Douglas and D. L. Hull, 180–238. Edinburgh: Edinburgh University Press.

Hacking, I. 1995. *Rewriting the Soul: Multiple Personality and the Sciences of Memory.* Princeton, NJ: Princeton University Press.

Halbwachs, M. 1968. *La mémoire collective.* Paris: Presses Universitaires de France.

Haraway, D. 1983. "Signs of Dominance: From a Physiology to a Cybernetics of Primate Society: C. R. Carpenter, 1930–1970." *Studies in the History of Biology* 6: 129–219.

Haraway, D. 1997. *Modest-Witness@Second-Millennium. FemaleMan-Meets-OncoMouse: Feminism and Technoscience.* New York: Routledge.

Hart, K. 1999. *The Memory Bank: Money in an Unequal World.* London: Profile Books.

Hartley, L. P. 1953. *The Go-between.* New York: Knopf.

Hartog, F. 1991. *Le miroir d'Hâerodote: Essai sur la reprâesentation de l'autre.* Paris: Gallimard.

Hawksworth, D. L. 1991. *Improving the Stability of Names: Needs and Options: Proceedings of an International Symposium, Kew, 20–23 February 1991.* Königstein/Taunus, Germany: Koeltz Scientific. Published for the International Association for Plant Taxonomy in conjunction with the International Union of Biological Sciences and the Systematics Association.

Hawksworth, D. L., and F. A. Bisby 1988. "Systematics: The Keystone of Biology." In *Prospects in Systematics*, edited by D. L. Hawksworth, 3–30. Oxford: Clarendon Press and Oxford University Press. Published for the Systematics Association.

Hawksworth, D. L., and R. K. Mibey 1997. "Information Needs of Inventory Programmes." In *Biodiversity Information: Needs and Options: Proceedings of the 1996 International Workshop on Biodiversity Information*, edited by D. L. Hawksworth, P. M. Kirk, and S. D. Clarke, 55–68. Oxon, UK, and New York: CAB International.

Hayden, C. P. 1998. "A Biodiversity Sampler for the Millenium." In *Reproducing Reproduction: Kinship, Power, and Technological Innovation*, edited by S. Franklin and H. Ragone, 173–206. Philadelphia: University of Pennsylvania Press.

Heims, S. J. 1975. "Encounter of Behavioral Sciences with New Machine-Organism Analogies in the 1940s." *Journal of the History of the Behavioral Sciences* 11: 368–373.

Heims, S. J. 1991. *The Cybernetics Group.* Cambridge, MA: MIT Press.

Helford, R. M. 1999. "Rediscovering the Presettlement Landscape: Making the Oak Savanna Ecosystem 'Real.'" *Science, Technology and Human Values* 24(1): 55–79.

Helmreich, S. 1998. *Silicon Second Nature Culturing Artificial Life in a Digital World.* Berkeley: University of California Press.

Heywood, V. H. 1991. "Needs for Stability of Nomenclature in Conservation." In *Improving the Stability of Names, Needs and Options: Proceedings of an International Symposium, Kew, 20–23 February 1991*, edited by D. L. Hawksworth, 53–58. Königstein/Taunus, Germany: Koeltz Scientific. Published for the International Association for Plant Taxonomy in conjunction with the International Union of Biological Sciences and the Systematics Association.

Heywood, V. H. 1997. "Information Needs in Biodiversity Assessments—From Genes to Ecosystems." In *Biodiversity Information: Needs and Options: Proceedings of the 1996 International Workshop on Biodiversity Information*, edited by D. L. Hawksworth, P. M. Kirk, and S. D. Clarke, 5–20. Oxon, UK, and New York: CAB International.

Heywood, V. H., et al. 1995. *Global Biodiversity Assessment.* Cambridge and New York: Cambridge University Press.

Hobsbawm, E. J. 1992. "Mass Producing Tradition: Europe, 1870–1914." In *The Invention of Tradition: Past and Present Publications*, edited by E. Hobsbawm and T. O. Ranger, 263–307. Cambridge and New York: Cambridge University Press.

Hobsbawm, E. J., and T. O. Ranger, eds. 1992. *The Invention of Tradition: Past and Present Publications.* Cambridge and New York: Cambridge University Press.

Holton, G. J. 1988. *Thematic Origins of Scientific Thought: Kepler to Einstein.* Cambridge, MA: Harvard University Press.

Huggett, R. J. 1997. *Environmental Change: The Evolving Ecosphere.* London: Routledge.

Hughes, N. F. 1991. "Improving Stability of Names: Earth Sciences Attitudes." In *Improving the Stability of Names: Needs and Options: Proceedings of an International Symposium, Kew, 20–23 February 1991*, edited by D. L. Hawksworth, 39–44. Königstein/Taunus, Germany: Koeltz Scientific. Published for the International Association for Plant Taxonomy in conjunction with the International Union of Biological Sciences and the Systematics Association.

Hull, D. L. 1966. "Phylogenetic Numericlature." *Systematic Zoology* 15: 14–17.

Humphries, C. J., et al. 1995. "Measuring Biodiversity Value for Conservation." *Annual Review of Ecology and Systematics* 26: 93–111.

Hunter, I. 1988. "Setting Limits to Culture." *New Formations* 4 (Spring): 103–123.

Huot, J-J. N. 1837. *Nouveau cours elémentaire de géologie.* Paris: Roret.

Hurst, C. J. 2000. *Viral Ecology.* San Diego and London: Academic.

Hutchins, E. 1995. *Cognition in the Wild.* Cambridge, MA: MIT Press.

ICSU-UNESCO. 1999. *World Conference on Science—Declaration on Science.* Paris: UNESCO.

Illingworth, V. 1983. *Oxford Dictionary of Computing.* Oxford and New York: Oxford University Press.

Ingersoll, R. C., et al. 1997. "A Model Information Management System for Ecological Research." *BioScience* 47(5): 310–316.

Jacker, C. 1964. *Man, Memory and Machines.* New York: Macmillan.

James, A. N., et al. 1999. "Balancing the Earth's Accounts." *Nature* 401: 323–324.

James, P. J. 1991. "A View from a Teacher." In *Improving the Stability of Names: Needs and Options: Proceedings of an International Symposium, Kew, 20–23 February 1991*, edited by D. L. Hawksworth, 59–65. Königstein/Taunus, Germany: Koeltz Scientific. Published for the International Association for Plant Taxonomy in conjunction with the International Union of Biological Sciences and the Systematics Association.

Jonasse, R. 2001. "Making Sense: Geographic Information Technologies and the Control of Heterogeneity." Ph.D. Thesis, Department of Communication, University of California at San Diego.

Journet, D. 1991. "Ecological Theories as Cultural Narratives: F. E. Clements's and H. A. Gleason's 'Stories' of Community Succession." *Written Communication* 8(4): 446–472.

Kevles, D. J. 1987. *The Physicists: The History of a Scientific Community in Modern America.* Cambridge, MA: Harvard University Press.

Khoshafian, S. 1993. *Object-Oriented Databases.* New York: Wiley.

Klemm, C. de. 1990. *Wild Plant Conservation and the Law.* Gland, Switzerland: IUCN-The World Conservation Union.

Koch, C. F. 1998. "'Taxonomic Barriers' and Other Distortions within the Fossil Record." In *The Adequacy of the Fossil Record,* edited by S. K. Donovan and C. R. C. Paul, 189–206. New York: Wiley.

Kuhn, T. S. 1970. *The Structure of Scientific Revolutions.* Chicago: University of Chicago.

Kurke, L. 1999. *Coins, Bodies, Games, and Gold: The Politics of Meaning in Archaic Greece.* Princeton, NJ: Princeton University Press.

Lakoff, G. 1987. *Women, Fire, and Dangerous Things: What Categories Reveal about the Mind.* Chicago: University of Chicago Press.

Lamb, H. H. 1995. *Climate, History, and the Modern World.* London and New York: Routledge.

Landor, W. S. 1824. *Imaginary Conversations of Literary Men and Statesmen.* London: Taylor and Hessey.

Laplace, P. S. 1799. *Exposition du Système du Monde.* Paris: Crapelet.

Laplace, P. S. 1814. *Essai philosophique sur les probabilités.* Paris: Courcier.

Latil, P. de. 1953. *Introduction à la cybernétique: La pensée artificielle* [Introduction to Cybernetics: Artificial Thought]. Paris: Gallimard.

Latour, B. 1987. *Science in Action: How to Follow Scientists and Engineers through Society.* Milton Keynes: Open University Press.

Latour, B. 1993a. *La clef de Berlin et autres leçons d'un amateur de sciences.* Paris: Éditions La Découverte.

Latour, B. 1993b. *We Have Never Been Modern.* Cambridge, MA: Harvard University Press.

Latour, B. 1996. "On Interobjectivity." *Mind, Culture and Activity: An International Journal* 3(4): 228–245.

Latour, B., and E. Hermant. 1998. *Paris ville invisible.* Paris: Les Empêcheurs de Penser en rond and Éditions La Découverte.

Le Goff, J. 1984. *The Birth of Purgatory.* Chicago: University of Chicago Press.

Le Goff, J. 1992. *History and Memory.* New York: Columbia University Press.

Leibniz, G. W., et al. 1994. *Writings on China.* Chicago, IL: Open Court.

Le Réformateur; journal quotidien des nouveaux intérêts matérials et moraux, industriels et politiques, littéraires et scientifiques. 1834–1835. Paris: 2 rue Dauphine.

Leroi-Gourhan, A. 1965. *Le geste et la parole: La mémoire et les rythmes.* Paris: Albin Michel.

Lettvin, J. Y., H. R. Matuana, W. S. McCulloch, and W. H. Pitts. 1959. "What the Frog's Eye Tells the Frog's Brain." *Proceedings of the Institute of Radio Engineers* 47: 1940–1951.

Levi, P. 2000. "The Ravine." In *The Oxford Book of Detective Stories,* edited by P. Craig, 312–328. Oxford: Oxford University Press.

Lévi-Strauss, C. 1971. *L'homme nu.* Paris, Plon.

Linde, C. 1993. *Life Stories: The Creation of Coherence.* New York: Oxford University Press.

Lipset, D. 1982. *Gregory Bateson: The Legacy of a Scientist.* Boston: Beacon Press.

Little, Frank J. 1964. "The Need for a Uniform System of Biological Numericlature." *Systematic Zoology* 13: 191–194.

Liu, A. Forthcoming. "Local Transcendence: Essays on Postmodern Historicism and the Database." Personal communication.

Lock, J. M. 1991. "The Index Kewensis and New Plant Names: Practical Considerations." In *Improving the Stability of Names: Needs and Options: Proceedings of an International Symposium, Kew, 20–23 February 1991*, edited by D. L. Hawksworth, 287–290. Königstein/Taunus, Germany: Koeltz Scientific. Published for the International Association for Plant Taxonomy in conjunction with the International Union of Biological Sciences and the Systematics Association.

Loewenberg, B. J. 1958. "Darwin, Darwinism and History." *General Systems—Yearbook of the Society for General Systems Research* 3: 7–17.

Loraux, N. 1996. *Né de la terre: mythe et politique à Athènes.* Paris: Seuil.

Lovejoy, A. O. 1936. *The Great Chain of Being; A Study of the History of an Idea. The William James Lectures Delivered at Harvard University, 1933.* Cambridge, MA: MIT Press.

Lucas, G. L. 1993. "A Worldwide Botanical Reference System." In *Designs for Global Plant Species Information Systems*, edited by F. A. Bisby, G. F. Russell, and R. J. Pankhurst, 9–12. Oxford: Clarendon Press and Oxford University Press. Published for the Systematics Association.

Lyell, C. 1830–1833. *Principles of Geology*, 3 vols. (vol. 1, 1830; vol. 2, 1832; vol. 3, 1833). London: J. Murray.

Lyell, C. 1833. *Lectures in Geology, King's College, London, Royal Institution, 1833 etc. Lyell Mss 7/8*, Edinburgh University, Edinburgh.

Lyell, Katherine M. 1881. *Life, Letters, and Journals of Sir Charles Lyell, Bart.* London: J. Murray.

Lyotard, J. F. 1984. *The Postmodern Condition: A Report on Knowledge.* Minneapolis: University of Minnesota Press.

Mabberley, D. J. 1991. The Problem of "Older" Names. In *Improving the Stability of Names: Needs and Options: Proceedings of an International Symposium, Kew, 20–23 February 1991*, edited by D. L. Hawksworth, 123–134. Königstein/Taunus, Germany: Koeltz Scientific. Published for the International Association for Plant Taxonomy in conjunction with the International Union of Biological Sciences and the Systematics Association.

MacKay, D. M. 1950. "Quantal Aspects of Scientific Information." *Philosophical Magazine* 41: 289–311.

MacKenzie, D. A. 1990. *Inventing Accuracy: An Historical Sociology of Nuclear Missile Guidance.* Cambridge, MA: MIT Press.

Mackintosh, T. S. 1840. *An Inquiry into the Nature of Responsibility as Deduced from Savage Justice, Civil Justice, and Social Justice: With some remarks upon the doctrine of irresponsibility, as taught by Jesus Christ and Robert Owen: also upon the responsibility of man to God.* Birmingham, West Midlands: J. Guest.

Maddison, W. P., and D. R. Maddison 1992. *MacClade: Analysis of Phylogeny and Character Evolution. Version 3.* Sunderland: Sinauer Associates.

Manovich, L. 2001. *The Language of New Media.* Cambridge, MA: MIT Press.

Mansell, R., et al. 1998. *Knowledge Societies: Information Technology for Sustainable Development.* Oxford and New York: Oxford University Press. Published for and on behalf of the United Nations.

Mansfield, M., A. Callahan, and L. J. Fogel. 1965. "Foreword." In *Biophysics and Cybernetic Systems: Proceedings of the Second Cybernetic Sciences Conference*, edited by M. Mansfield, A. Callahan, and L. J. Fogel, i–xv. Washington, DC: Spartan Books.

Markus, L. F. 1993. "The Goals and Methods of Systematic Biology." In *Advances in Computer Methods for Systematic Biology: Artificial Intelligence, Databases, Computer Vision*, edited by R. Fortuner, 16–49. Baltimore, MD: Johns Hopkins University Press.

Masani, P. R. 1990. *Norbert Wiener: 1894–1964.* Basel: Birkhauser Verlag.

Matsuda, M. 1996. *The Memory of the Modern.* New York: Oxford University Press.

Matthews, R. E. F. 1983. *A Critical Appraisal of Viral Taxonomy.* Boca Raton, FL: CRC Press.

Mayr, E. 1988. "Recent Historical Developments." In *Prospects in Systematics*, edited by D. L. Hawksworth, 31–43. Oxford: Clarendon Press and Oxford University Press. Published for the Systematics Association.

McLuhan, M., and B. R. Powers. 1975. *The Global Village.* New York: Oxford University Press.

McLuhan, M., and Q. Fiore. 1996. *The Medium Is the Massage: An Inventory of Effects.* San Francisco: HardWired.

McNeill, J. 1991. "Instability in Biological Nomenclature: Problems and Solutions." In *Improving the Stability of Names: Needs and Options: Proceedings of an International Symposium, Kew, 20–23 February 1991*, edited by D. L. Hawksworth, 94–108. Königstein/Taunus, Germany: Koeltz Scientific. Published for the International Association for Plant Taxonomy in conjunction with the International Union of Biological Sciences and the Systematics Association.

Melville, H. [1852] 1995. *Pierre or The Ambiguities: The Kraken Edition.* New York: HarperCollins.

Michelet, J. 1959. *Journal. Texte intégral, établi sur les manuscrits autographes et publié pour la première fois, avec une introd., des notes et de nombreux documents inédits.* Paris: Gallimard.

Michelet, J. [1828–1831] 1971. "Introduction à l'histoire universelle." In *Oeuvres complètes*, vol. 2, edited by P. Viallaneix, v. Paris: Flammarion.

Michener, W. K., et al. 1997. "Nongeospatial Metadata for the Ecological Sciences." *Ecological Applications* 7(1): 330–342.

Michener, W. K., et al., eds. 1998. *Data and Information Management in the Ecological Sciences: A Resource Guide.* Albuquerque, NM: LTER Network Office, University of New Mexico.

Miller, A. G. 1991. "Transformations of Time and Space; Oaxaca, Mexico, circa 1500–1700." In *Images of Memory: On Remembering and Representation*, edited by W. S. Melion and S. Küchler, 141–175. Washington, DC: Smithsonian Institution Press.

Mosco, V. 2004. *The Digital Sublime: Myth, Power, and Cyberspace.* Cambridge: MA: MIT Press.

Murphy, F. A., et al., eds. 1995. *Virus Taxonomy: Classification and Nomenclature of Viruses. Sixth Report of the International Committee on Taxonomy of Viruses. Virology Division International Union of Microbiological Societies.* Vienna: Springer-Verlag.

Muses, C. 1962. "The Logic of Biosimulation." In *Aspects of the Theory of Artificial Intelligence: The Proceedings of the First International Symposium on Biosimulation, Locarno, June 29–July 5, 1960*, edited by C. A. Muses, 115–163. New York: Plenum.

Museum of Natural History. 2002. "Measuring Biodiversity Value." http://www.nhm.ac.uk/science/projects/worldmap/diversity/index.html.

Myers, N. 1997. "Biodiversity's Genetic Library." In *Nature's Services: Societal Dependence on Natural Ecosystems*, edited by G. C. Daily, 255–273. Washington, DC: Island Press.

Needham, J. G. 1910. "Practical Nomenclature." *Science* 213: 295–300.

Neisser, U. 1966. "Computers as Tools and Metaphors." In *The Social Importance of Cybernetics*, edited by C. R. Dechert, 206–217. New York: Simon and Schuster.

Newell, A. 1983. "Some Intellectual Issues in the History of Artificial Intelligence." In *The Study of Information: Interdisciplinary Messages*, edited by F. Machlup and U. Mansfield, 187–227. New York: Wiley.

Nietzsche, F. 1957. *The Use and Abuse of History.* Indianapolis: The Bobbs-Merrill Company.

Nolin, O. 1998. *Bonaparte et les savants français en Egypte: 1798–1801.* Paris: Ed. Mille et une nuits.

O'Hara, R. J. 1992. "Telling the Tree: Narrative Representation and the Study of Evolutionary History." *Biology and Philosophy* 7: 135–160.

O'Malley, M. 1996. *Keeping Watch: A History of American Time.* Washington, DC: Smithsonian Institution Press.

O'Neill, R. V. 2001. "Is It Time to Bury the Ecosystem Concept? (With Full Military Honors, Of Course!)." *Ecology* 82(12): 3275–3284.

Padel, F. 1998. "Forest Knowledge: Tribal People, Their Environment and the Structure of Power." In *Nature and the Orient: The Environmental History of South and Southeast Asia*, edited by R. Grove, V. Damodaran, and S. Sangwan, 891–917. Delhi and New York: Oxford University Press.

Paley, W. 1802. *Natural Theology; or, Evidences of the Existence and Attributes of the Deity. Collected from the Appearances of Nature.* London: R. Faulder.

Panchen, A. L. 1992. *Classification, Evolution, and the Nature of Biology.* Cambridge and New York: Cambridge University Press.

Pankhurst, R. 1993. "Taxonomic Databases: The PANDORA System." In *Advances in Computer Methods for Systematic Biology: Artificial Intelligence, Databases, Computer Vision*, edited by R. Fortuner, 229–235. Baltimore, MD: Johns Hopkins University Press.

Pask, G. 1961a. *An Approach to Cybernetics.* London: Hutchinson.

Pask, G. 1961b. "A Proposed Evolutionary Model." In *Principles of Self-organization*, edited by H. von Foerster and G. Zopf, 229–254. London: Pergamon Press.

Patterson, D. J. and J. Larsen 1991. "Nomenclatural Problems with Protists." In *Improving the Stability of Names: Needs and Options: Proceedings of an International Symposium, Kew, 20–23 February 1991*, edited by D. L. Hawksworth, 197–208. Königstein/Taunus, Germany: Koeltz Scientific. Published for the International Association for Plant Taxonomy in conjunction with the International Union of Biological Sciences and the Systematics Association.

Paul, C. R. C. 1998. "Adequacy, Completeness and the Fossil Record." In *The Adequacy of the Fossil Record*, edited by S. K. Donovan and C. R. C. Paul, 1–22. Chichester and New York: Wiley.

Pennisi, E. 1999. "Did Cooked Tubers Spur the Evolution of Big Brains?" *Science* 283 (March 26): 2004–2005.

Perron, C. de. [1835] 1840. *Système complètement neuf de classification du règne animal ramenant celle-ci aux seuls véritables principes qui puissent lui servir de base.* Paris: Pourreau.

Peterson, J. S. 1993. "U.S. Interagency Botanical Data Applications, Needs and the PLANTS Database." In *Designs for Global Plant Species Information Systems*, edited by F. A. Bisby, G. F. Russell, and R. J. Pankhurst, 13–19. Oxford: Clarendon Press and Oxford University Press. Published for the Systematics Association Press.

Pfeiffer, J. 1962. *The Thinking Machine.* New York: Lippincott.

Poincaré, H. 1905. *Science and Hypothesis.* New York: The Science Press.

Poinsot, L. 1837. *Eléments de Statique, suivis de trois mémoires.* Paris: Bachelier.

Porter, R. 1977. *The Making of Geology: Earth Science in Britain, 1660–1815.* Cambridge and New York: Cambridge University Press.

Porter, T. M. 1986. *The Rise of Statistical Thinking, 1820–1900.* Princeton, NJ: Princeton University Press.

Posey, D. A. 1997. "Wider User and Application of Indigenous Knowledge, Innovations and Practices: Informations Systems and Ethical Concerns." In *Biodiversity Information, Needs and Options: Proceedings of the 1996 International Workshop on Biodiversity Information*, edited by D. L. Hawksworth, P. M. Kirk, and S. D. Clarke, 69–103. Oxon, UK, and New York: CAB International.

President's Committee of Advisors on Science and Technology. 1998. "Teaming with Life: Investing in Science to Understand and Use America's Living Capital." Washington PCAST Panel on Biodiversity and Systems.

Proust, M. 1989. *A la recherche du temps perdu, tome iv.* Paris: Pleiade.

Prout, W. 1834. *Chemistry, Meteorology, and the Function of Digestion, Considered with Reference to Natural theology.* Philadelphia: Carey Lea and Blanchard.

Pynchon, T. 1997. *Mason & Dixon.* New York: Holt.

Pyne, S. J. 1991. *Burning Bush: A Fire History of Australia.* New York: Holt.

Pyne, S. J. 1997. *Fire in America: A Cultural History of Wildland and Rural Fire.* Seattle: University of Washington Press.

Qian, Sima, Ssuma Ch'ien. 1994. *Historical Records/ Sima Qian; Translated with an Introduction and Notes by Raymond Dawson.* Oxford: Oxford University Press.

Quastler, H. 1964. *The Emergence of Biological Organization.* New Haven: Yale University Press.

Rabelais, F. 1995. *Oeuvres conplètes.* Paris: Sevil.

Raven, P. H., et al. 1971. "The Origins of Taxonomy: A Review of Its Historical Development Shows Why Taxonomy Is Unable to Do What We Expect of It." *Science* 174 (December 17): 1210–1213.

Rayward, W. B. 1975. *The Universe of Information: The Work of Paul Otlet for Documentation and International Organisation.* Moscow: All-Union Institute for Scientific and Technical Information (VINITI). Published for International Federation for Documentation (FID).

Read, H. E. 1968. *The Meaning of Art.* London: Faber.

Reardon, J. 2001. "The Human Genome Diversity Project: A Case Study in Coproduction." *Social Studies of Science* 31(3): 357–388.

Renne, P. R., D. B. Karner, and K. R. Ludwig. 1998. "Absolute Ages Aren't Exactly." *Science* 282 (December 4): 1840–1841.

Reveal, J. L. 2004. "A Nomenclatural Morass: Dramatis Personae—Menzies, Lambert and Poiret." http://www.lewis-clark.org/content/content-article.asp?ArticleID=1509.

Reynaud, J. 1833. "De l'infinité du ciel." *Revue Encyclopédique* 58: 1–20.

Rich, A. 1978. "Cartographies of Silence." In *The Dream of a Common Language: Poems, 1974–1977,* A. Rich. New York: W. W. Norton.

Richards, T. 1996. *The Imperial Archive: Knowledge and the Fantasy of Empire.* London: Verso.

Ride, W. D. L. 1991. "Justice for the Living: A Review of Bacteriological and Zoological Initiatives in Nomenclature." In *Improving the Stability of Names: Needs and Options: Pro-*

ceedings of an International Symposium, Kew, 20–23 February 1991, edited by D. L. Hawksworth, 105–122. Königstein/Taunus, Germany: Koeltz Scientific. Published for the International Association for Plant Taxonomy in conjunction with the International Union of Biological Sciences and the Systematics Association.

Ridley, M. 1986. *Evolution and Classification; The Reformation of Cladism.* London: Longman.

Roberts, M. N., et al. 1996. *Memory: Luba Art and the Making of History.* New York: Museum for African Art.

Roblin, R. 1997. "Resources for Biodiversity in Living Collections and the Challenges of Assessing Microbial Biodiversity." In *Biodiversity II: Understanding and Protecting our Biological Resources*, edited by M. L. Reaka-Kudla, D. E. Wilson, and E. Wilson, 467–474. Washington, DC: Joseph Henry Press:

Rosenberg, C. E. 1976. *No Other Gods: on Science and American Social thought.* Baltimore, MD: Johns Hopkins University Press.

Rosenblueth, A., N. Wiener, and J. Bigelow. 1943. "Behavior, Purpose and Teleology." *Philosophy of Science* 10: 18–24.

Roy, K., and M. Foote 1997. "Morphological Approaches to Measuring Biodiversity." *TREE (Trends in Ecology and Evolution)* 12(7): 277–281.

Saley, H. E. 1960. "Air Force Research on Living Prototypes." In *Bionics Symposium*, 41–43. Dayton, OH: Wright Air Development Division.

Schachter, D. L. 1996. *Searching for Memory: the Brain, the Mind, and the Past.* New York: Basic Books.

Schama, S. 1995. *Landscape and Memory.* New York: Knopf. Distributed by Random House.

Schivelbusch, W. 1986. *The Railway Journey: The Industrialization of Time and Space in the Nineteenth Century.* Berkeley: University of California Press.

Schmandt-Besserat, D. 1992. *Before Writing.* Austin: University of Texas Press.

Scott-Ram, N. 1990. *Transformed Cladistics, Taxonomy and Evolution.* Cambridge: Cambridge University Press.

Scrope, G. P. 1833. *Principles of Political Economy, Deduced from the Natural Laws of Social Welfare, and Applied to the Present State of Britain.* London: Longman, Rees, Orme, Brown and Green.

Sebald, W. G. 2002. *After Nature.* London: Hamish Hamilton.

Serres, M. 1991. *Le tiers instruit.* Paris: Flammarion.

Serres, M. 1980. *Le passage du nord-ouest.* Paris: Éditions de Minuit.

Serres, M. 1982a. *Genèse.* Paris: B. Grasset.

Serres, M. 1982b. *The Parasite.* Baltimore, MD: Johns Hopkins University Press.

Serres, M. 1987. *Statues: Le second livre des fondations.* Paris: Éditions François Bourin.

Serres, M. 1990. *Le contrat naturel.* Paris: Éditions François Bourin.

Serres, M. 1993. *Les origines de la géométrie.* Paris: Flammarion.

Serres, M. 2001. *Hominescence.* Paris: Pommier.

Shapin, S., and S. Schaffer. 1989. *Leviathan and the Air-Pump: Hobbes, Boyle, and the Experimental Life.* Princeton: Princeton University Press.

Sissa, G., and M. Détienne 1989. *La vie quotidienne des dieux grecs.* Paris: Hachette.

Smet, W. M. A. de 1991. "Meeting User Needs by an Alternative Nomenclature." In *Improving the Stability of Names: Needs and Options: Proceedings of an International Symposium, Kew, 20–23 February 1991*, edited by D. L. Hawksworth, 179–181. Königstein/Taunus, Germany: Koeltz Scientific. Published for the International Association for Plant Taxonomy in conjunction with the International Union of Biological Sciences and the Systematics Association.

Sohn-Rethel, A. 1975. "Science as Alienated Consciousness." *Radical Science Journal* 5: 65–101.

Sohn-Rethel, A. 1978. *Intellectual and Manual Labour: A Critique of Epistemology.* London: Macmillan.

Stace, C. A. 1991. "Why the Complaints Are Justified, or Don't You Have Something Better to Do?" In *Improving the Stability of Names: Needs and Options: Proceedings of an International Symposium, Kew, 20–23 February 1991*, edited by D. L. Hawksworth, 72–78. Königstein/Taunus, Germany: Koeltz Scientific. Published for the International Association for Plant Taxonomy in conjunction with the International Union of Biological Sciences and the Systematics Association.

Star, S. L. 1991. "The Sociology of the Invisible: The Primacy of Work in the Writings of Anselm Strauss." In *Social Organization and Social Process: Essays in Honor of Anselm Strauss*, ed. D. Maines, 265–283. Hawthorne, NY: Aldine de Gruyter.

Star, S. L., and K. Ruhleder. 1996. "Steps toward an Ecology of Infrastructure: Design and Access for Large Information Spaces." *Information Systems Research* 7(1): 111–134.

Stevens, P. F. 1991. "George Bentham and the Kew Rule." In *Improving the Stability of Names: Needs and Options: Proceedings of an International Symposium, Kew, 20–23 February 1991*, edited by D. L. Hawksworth, 157–168. Königstein/Taunus, Germany: Koeltz Scientific. Published for the International Association for Plant Taxonomy in conjunction with the International Union of Biological Sciences and the Systematics Association.

Stevens, P. F. 1997. "J. D. Hooker, George Bentham, Asa Gray and Ferdinand Mueller on Species Limits in Theory and Practice: A Mid-nineteenth-century Debate and Its Repercussions." In *Historical Records of Australian Science* 11(3): 345–370.

Stevens, P. F., and S. P. Cullen 1990. "Linnaeus, the Cortex-Medulla Theory, and the Key to His Understanding of Plant Form and Natural Relations." *Journal of the Arnold Arboretum* 71(April): 179–220.

Stork, N. E. 1997. "Measuring Global Biodiversity and Its Decline." In *Biodiversity II: Understanding and Protecting our Biological Resources*, edited by M. L. Reaka-Kudla, D. E. Wilson, and E. Wilson, 41–68. Washington, DC: Joseph Henry Press.

Strauss, E. 1999. "Can Mitochondrial Clocks Keep Time?" *Science* 283: 1435–1438.

Sue, E. 1844. *Le juif errant* [The Wandering Jew]. New York: Gaillardet.

Takacs, D. 1996. *The Idea of Biodiversity: Philosophies of Paradise.* Baltimore, MD: Johns Hopkins University Press.

Tanaka, S. 2004. *New Times in Modern Japan.* Princeton: Princeton University Press.

Tanenbaum, A. S. 1996. *Computer Networks.* Upper Saddle River, NJ: Prentice Hall PTR.

Taxacom Discussion Listserv. Various dates. http://listserv.nhm.ku.edu/archives/taxacom.html.

Tisseron, S. 1996. *Le mystère de la chambre claire.* Paris: Flammarion.

Toledo Maya Cultural Council and Toledo Alcaldes Association. 1997. *Maya Atlas: The Struggle to Preserve Maya Land in Southern Belize.* Berkeley and Emeryville, CA: North Atlantic Books. Distributed by Publishers Group West.

Tort, P. 1989. *La Raison Classificatoire: les Complexes Discursifs—Quinze Etudes.* Paris: Aubier.

Turnbull, D. 2000. *Masons, Tricksters, and Cartographers: Comparative Studies in the Sociology of Scientific and Indigenous Knowledge.* Australia: Harwood Academic.

Vane-Wright, R. I., et al. 1991. "What to Protect?—Systematics and the Agony of Choice." *Biological Conservation* 55: 235–254.

Vernon, K. 1993. "Desperately Seeking Status: Evolutionary Systematics and the Taxonomists' Search for Respectability 1940–1960." *British Journal for the History of Science* 26: 207–227.

Veyne, P. 1971. *Comment on écrit l'histoire; augmenté de Foucault révolutionne l'histoire.* Paris: Éditions du Seuil.

Vrba, E. S. 1994. "An Hypothesis of Heterochrony in Response to Climatic Cooling and Its Relevance to Early Hominid Evolution." In *Integrative Paths to the Past: Paleoanthropological Advances in Honor of F. Clark Howell*, edited by R. L. Ciochon and R. S. Corruccini, 345–376. Englewood Cliffs, NJ: Prentice Hall.

Walsh, J. P., and G. R. Ungson. 1991. "Organizational Memory." *Academy of Management Review* 16(1): 57–91.

Walters, S. M. 1986. "The Name of the Rose: A Review of Ideas on the European Bias in Angiosperm Classification." *The New Phytologist* 104: 527–546.

Watson-Verran, H., and D. Turnbull 1995. "Science and Other Indigenous Knowledge Systems." In *Handbook of Science and Technology Studies*, edited by S. Jasanoff, G. Markle, J. Petersen, and T. Pinch, xv, 820. Thousand Oaks, CA: Sage Publications.

Weart, S. 1988. *Nuclear Fear: A History of Images.* Cambridge, MA: Harvard University Press.

Weedman, J. 1998. "The Structure of Incentive: Design and Client Roles in Application-Oriented Research." *Science, Technology, & Human Values* 23(3): 315–345.

Weinberg, S. 1993. *The First Three Minutes: A Modern View of the Origin of the Universe.* New York: Basic Books.

Wettstein, R. von and J., ed. 1906. *Actes du Congrès International de Botanique tenu à Vienne (Autriche 1905).* Jena: Verlag von Gustav Fischer.

Wheeler, Q. D., and J. Cracraft 1997. "Taxonomic Preparedness: Are We Ready to Meet the Biodiversity Challenge?" In *Biodiversity II: Understanding and Protecting Our Biological Resources*, edited by M. L. Reaka-Kudla, D. E. Wilson, and E. Wilson, 435–446. Washington, DC: Joseph Henry Press.

Whewell, W. 1833. *Astronomy and General Physics Considered with Reference to Natural Theology.* Philadelphia: Carey Lea and Blanchard.

Wiener, N. 1948. *Cybernetics; or Control and Communication in Man and Machine.* New York: Wiley.

Wiener, N. 1951. *Cybernetics and Society.* New York: Executive Techniques.

Wiener, N. 1956. *I Am a Mathematician: The Later Life of an Ex-prodigy.* Garden City, NY: Doubleday.

Wiener, N. 1964. *God and Golem, Inc.: A Comment on Certain Points Where Cybernetics Impinges on Religion.* London: Chapman and Hall.

Wiesner, J. von, ed. 1906. *Actes due Congrès International de Botanique tenu à Vienne (Autriche 1905).* Jena: Verlag von Gustav Fischer.

Williams, P. H., et al. 1994. "Do Conservationists and Molecular Biologists Value Differences between Organisms in the Same Way?" *Biodiversity Letters* 2(3): 67–78.

Williams, P., et al. 1997. "Descriptive and Predictive Approaches to Biodiversity Measurement." *TREE* 12(11): 444–445.

Williams, R. 1983. *Keywords: A Vocabulary of Culture and Society.* London: Fontana Paperbacks.

Wilson, E. O., et al. 1988. *Biodiversity.* Washington, DC: National Academy Press.

Wilson, R. A. 1999. *Species: New Interdisciplinary Essays.* Cambridge, MA: MIT Press.

Winchester, S. 2001. *The Map that Changed the World: William Smith and the Birth of Modern Geology.* New York: HarperCollins.

Wing, S. L. 1997. "Global Warming and Plant Species Richness: A Case Study of the Paleocene/Eocene Boundary." In *Biodiversity II: Understanding and Protecting Our Biological Resources*, edited by M. L. Reaka-Kudla, D. E. Wilson, and E. Wilson, 163–185. Washington, DC: Joseph Henry Press.

Winston, J. E. 1992. "Systematics and Marine Conservation." In *Systematics, Ecology, and the Biodiversity Crisis*, edited by N. Eldredge, 144–168. New York: Columbia University Press.

Wittezaele, J. J., and T. Garcia. 1992. *A la recherche de l'école de Palo Alto.* Paris: Seuil.

Wolfram, H. 1988. *History of the Goths.* Berkeley: University of California Press.

Woolgar, S. 1987. "Configuring the User: The Case of Usability Trials." In *A Sociology of Monsters: Essays on Power, Technology and Domination*, edited by J. Law, 58–97. London: Routledge.

Worster, D. 1994. *Nature's Economy: A History of Ecological Ideas.* Cambridge and New York: Cambridge University Press.

Yates, F. 1966. *The Art of Memory.* Chicago: University of Chicago Press.

Yates, J. 1989. *Control through Communication: The Rise of System in American Management.* Baltimore, MD: Johns Hopkins University Press.

Yates, J., and W. J. Orlikowski 1992. "Genres of Organizational Communication: A Structurational Approach to Studying Communication and Media." *Academy of Management Review* 17: 299–326.

Yeates, D. K. 1995. "Groundplans and Exemplars: Paths to the Tree of Life." *Cladistics* 11: 343–357.

Yerushalmi, Y. H. 1988. *Usages de l'oubli.* Paris: Seuil.

Yerushalmi, Y. H. 1996. *Zakhor: Jewish History and Jewish Memory.* Seattle: University of Washington Press.

Young, R. 1973. "The Historiographic and Ideological Contexts of the Nineteenth-Century Debate on Man's Place in Nature." In *Changing Perspectives in the History of Science: Essays in Honour of Joseph Needham*, edited by J. Needham, M. Teich, and R. M. Young, 110–187. London: Heinemann Educational.

Žižek, S. 2000. *The Fragile Absolute, or, Why Is the Christian Legacy Worth Fighting For?* London and New York: Verso.

Inside Technology

edited by Wiebe E. Bijker, W. Bernard Carlson, and Trevor Pinch

Janet Abbate, *Inventing the Internet*

Charles Bazerman, *The Languages of Edison's Light*

Marc Berg, *Rationalizing Medical Work: Decision Support Techniques and Medical Practices*

Wiebe E. Bijker, *Of Bicycles, Bakelites, and Bulbs: Toward a Theory of Sociotechnical Change*

Wiebe E. Bijker and John Law, editors, *Shaping Technology/Building Society: Studies in Sociotechnical Change*

Stuart S. Blume, *Insight and Industry: On the Dynamics of Technological Change in Medicine*

Pablo J. Boczkowski, *Digitizing the News: Innovation in Online Newspapers*

Geoffrey C. Bowker, *Memory Practices in the Sciences*

Geoffrey C. Bowker, *Science on the Run: Information Management and Industrial Geophysics at Schlumberger, 1920–1940*

Geoffrey C. Bowker and Susan Leigh Star, *Sorting Things Out: Classification and Its Consequences*

Louis L. Bucciarelli, *Designing Engineers*

H. M. Collins, *Artificial Experts: Social Knowledge and Intelligent Machines*

Paul N. Edwards, *The Closed World: Computers and the Politics of Discourse in Cold War America*

Herbert Gottweis, *Governing Molecules: The Discursive Politics of Genetic Engineering in Europe and the United States*

Gabrielle Hecht, *The Radiance of France: Nuclear Power and National Identity after World War II*

Kathryn Henderson, *On Line and On Paper: Visual Representations, Visual Culture, and Computer Graphics in Design Engineering*

Anique Hommels, *Unbuilding Cities: Obduracy in Urban Sociotechnical Change*

David Kaiser, editor, *Pedagogy and the Practice of Science: Historical and Contemporary Perspectives*

Peter Keating and Alberto Cambrosio, *Biomedical Platforms: Reproducing the Normal and the Pathological in Late-Twentieth-Century Medicine*

Eda Kranakis, *Constructing a Bridge: An Exploration of Engineering Culture, Design, and Research in Nineteenth-Century France and America*

Pamela E. Mack, *Viewing the Earth: The Social Construction of the Landsat Satellite System*

Donald MacKenzie, *Inventing Accuracy: A Historical Sociology of Nuclear Missile Guidance*

Donald MacKenzie, *Knowing Machines: Essays on Technical Change*

Donald MacKenzie, *Mechanizing Proof: Computing, Risk, and Trust*

Maggie Mort, *Building the Trident Network: A Study of the Enrolment of People, Knowledge, and Machines*

Nelly Oudshoorn and Trevor Pinch, editors, *How Users Matter: The Co-Construction of Users and Technologies*

Paul Rosen, *Framing Production: Technology, Culture, and Change in the British Bicycle Industry*

Susanne K. Schmidt and Raymund Werle, *Coordinating Technology: Studies in the International Standardization of Telecommunications*

Charis Thompson, *Making Parents: The Ontological Choreography of Reproductive Technology*

Dominque Vinck, editor, *Everyday Engineering: An Ethnography of Design and Innovation*

Index

Printed in the United States
by Baker & Taylor Publisher Services

Printed in the United States
by Baker & Taylor Publisher Services